AF539943

ENCYCLOPEDIA OF POLYMER SCIENCE AND TECHNOLOGY

Volume - 3

ENCYCLOPEDIA OF POLYMER SCIENCE AND TECHNOLOGY

Volume - 3

by

Navid Naderpour
Bandar Imam Petrochemical Complex BIPC
Tarbiat Modares University of Tehran
Engineering Faculty-Polymer Department
Research & Development center of BIPC
Islamic Republic of Iran

Ebrahim Vasheghani Farahani
Tarbiat Modares University of Tehran
Engineering Faculty-Polymer Department
Islamic Republic of Iran

Adel Nejad Salim
National Petrochemical Company NPC
Islamic Republic of Iran

Reza Amiri
Bandar Imam Petrochemical Complex - BIPC
Islamic Republic of Iran

Simin Eydivand
Bandar Imam Petrochemical Complex BIPC
Research & Development center of BIPC
Islamic Republic of Iran

2009

SBS Publishers & Distributors Pvt. Ltd.
New Delhi

ISBN SET - 9788189741921
ISBN V-3 - 9788189741950

First Published in India in 2009

Published by:
SBS PUBLISHERS & DISTRIBUTORS PVT. LTD.
2/9, Ground Floor, Ansari Road, Darya Ganj,
New Delhi - 110002, INDIA
Tel: 23289119, 41563911
Email: mail@sbspublishers.com

Printed in India by Syndicate Binders, Noida.

Special Thanks to

Mr. KIYANOSH SHAMSEAMIRI - Manager, Basparan Complex
Mr. FATHE - Education Manager, NPC
Mr. FAKKARI - Head, Education Center, BIPC
And Specially Dr Nader Nadepour
for encouragement, help and advice.

We would also like to thank Dr HADADI ASL, Dr NASR JAFARI and Dr GHAFELEBASHI of Petrochemical Research and Technology Co. (NPC-RT).

We also thank
Dr M.NAVIDFAMILI,
Dr M.SEMSARZADEH,
& Dr M KOKABI
of Polymer Department of TARBIAT MODARES University

Mr. AKBARPOUR,
M.GHOLAMZADEH,
& D.MOHAMADI
of Basparan Engineering Complex

Mr. H.GODARZBOOD,
M.JOHARI,
M.HAMEDIAN,
A.MOAVI
& R.ADIBAN,
Researchers of R & D Center, BIPC
for providing us the visuals and figures for this Encyclopedia.

And finally we thank many people who have helped, often without knowing they were doing so, for, exhortation and consolation throughout the years it has taken to create this work.

Preface

This publication entitled "Encyclopedia of Polymer Science and Technology" provides the readers with key information on the said subject in an affordable three volumes condensed format. This user-friendly reference publication offers quick access to all areas of polymer science and technology. It presents the state of the art in all areas of polymer science, technology and engineering, covering new imaging and analytical techniques and new methods of controlled polymer architecture. It also talks about related information about polymers, plastics, fibers, biomaterials and polymerization processes.

It follows a distinctly new approach while discussing the interdisciplinary nature of polymer science, technology and engineering. Modern polymer science, technology and engineering are firmly rooted both in the chemistry of macromolecules and their physical chemistry. Besides, this publication provides readers with information on all the important uses of synthetic polymers. This publication fulfills the much desired need for such compact volumes on the said subject. The elaborate bibliography/reference to all applied aspects makes this book an indispensable support for both students and professionals.

This "Encyclopedia of Polymer Science and Technology" gives an introduction to polymer science and technology in addition to a basic understanding of polymers and polymer nomenclature and polymer characterization. Natural polymers, synthetic polymers and their sources are discussed briefly. Other themes which are covered include the phenomena of observing and measuring polymer/colloids properties; polymer chemistry and chemical properties of polymers; chemistry of high polymers; polymer physics and physical properties of polymers; polymer structure, tacticity and other properties; and polymer measurement devices. This publication also addresses processes like methods of polymer manufacturing; methods of polymer synthesis and modification; design and synthesis of polymers with controlled structures; products and stuffs made from polymers; polymer processing, testing and analysis; polymer blends and composites; polymer fibre composite materials; design and manufacture of composites; biopolymers and biomaterials; smart polymers and polymeric materials; and polymers strength enhancement and degradation. Other aspects which are highlighted include computational polymer science and technology; theoretical polymer science and technology; applied polymer science and technology; polymer Rheology; polymer crystallization; polymer material science and engineering; biomedical applications of polymers; macromolecules of interfaces and films; electronic and photonic molecular materials and devices; advanced polymer research; and polymer fibre applications.

The content of this publication on polymer science and technology is aimed for qualified and unqualified graduates who possess knowledge on polymer chemistry, polymer physics, polymer technology, polymer engineering and their applications. It also provides such students a skill on practical works related to polymer chemistry and processing. The objectives of this publication are to provide a broad range of knowledge in polymer science and technology as well as the skill/ability to apply this knowledge for industrial processes and modification of polymer products. This will also enhance opportunity of the graduate students for further education in polymer science and technology or related fields.

Contents

1

Computational Polymer Science and Technology

1.1 Computational Design of Imprinted Polymers

The process we have developed at Cranfield involves modelling the formation of the complex between template and functional monomers present in the pre-polymerisation mixture.

The computational approach permits:

(*a*) Selection of the best monomers for polymer formation, i.e., those possessing highest affinity for the template.
(*b*) Mimicking of the specific polymerisation or binding conditions by modifying the dielectric constant of the medium and the atomic charges within the monomer/template models.
(*c*) Prediction of the relative specificity and affinity of the polymer.

MIP Design Protocol

The Template

A molecular model of the template is made, the charges for each atom are calculated and the structure of the template is refined using molecular mechanics.

The Monomer Database

A functional monomer database is constructed for screening against the template. Libraries contain from 21 to 260 functional monomer structures comprising acidic, basic and neutral molecules, the former being represented both in their ionised and unionised states.

Screening

Each of the members of the virtual library is then probed for its possible interactions with the template molecule using the Leapfrog™ algorithm. Leapfrog™ is a component of the SYBYL™ software package (Tripos) and is a second-generation *de novo* drug discovery program that allows for the evaluation of potential ligand structures. The results for each monomer are compared on the basis of their binding scores, resulting in a table ranking the monomers according to their binding scores, expressed as an energy value (in kcal/mol). Monomers at the top of this table are therefore judged to be the best candidates for polymer preparation.

Refining the Design

Although the Leapfrog™ algorithm gives an indication as to which monomers may be best for polymer preparation, a refining step, simulated annealing, is carried out in order to allow us to gather more information before polymer preparation. Simulated annealing is a general-purpose global optimisation technique based on a constrained molecular dynamics simulation, which can therefore be used to optimise the monomer/template composition based on the screening results. This refining step determines the type and quantity of monomer units, which may be used for polymer preparation. Molecular mechanics/ dynamics is used to simulate the pre-arrangement of the functional monomers with the template, as it exists in the monomer mixture prior to polymerisation. The goal of the refining step is to extract information about the exact stoichiometric ratios of monomers to template. The computer modelling also allows us the flexibility to change parameters applied to the refining step such as the dielectric constant of the medium and the temperature chosen for the annealing process in order to mimic the conditions (including solvent) encountered in the actual real polymerisation or re-binding experiments.

Synthesis and Testing

The computationally designed polymer is then synthesised and tested. An analysis of the affinity and selectivity of computationally designed MIPs shows that they possess similar properties to antibodies raised against the same molecule.

Table 1.1: Affinity and Sensitivity Range of MIPs and Antibodies—Comparison Study.

Receptor	*Kd, nM*	*Sensitivity Range (μg l^{-1})*
Computational MIP	0.3 ± 0.08	0.1-100
Monoclonal antibody	0.03 ± 0.004	0.025-5
Polyclonal antibody	0.5 ± 0.07	0.05-10

1.2 Research Group on Computational Polymer Science

Activity

Computational simulation is widespread across a variety of research and development fields in both industry and academia. Although polymer science consists of a vast research area; covering, e.g. chemical reactions such as catalytically, thermally, or radiatively induced polymerization, cross-linking, and decomposition, structure and morphology of a single chain, crystal, phase separation, and micelles, properties of functional polymers such as thermodynamic properties, electronic properties, mechanical strength, rheological behavior, solubility and diffusivity of other molecules, as well as their applications; there is no place computational simulation is not related to. Our research group, therefore, is acting to serve in three important subjects. *Education:* We provide an annual lecture mainly on the basic of computational science for young researchers and beginners. *Research:* We hold a collaborative research meeting with other research groups so as to provide an opportunity for computational and experimental chemists to exchange their opinions and stimulate each other. *Communication:* At our one-night summer camp, young researchers and experts can discuss literally with no time limits about their interests and thoughts. We are looking forward to seeing you in our events.

1.3 Computational Polymer Physics

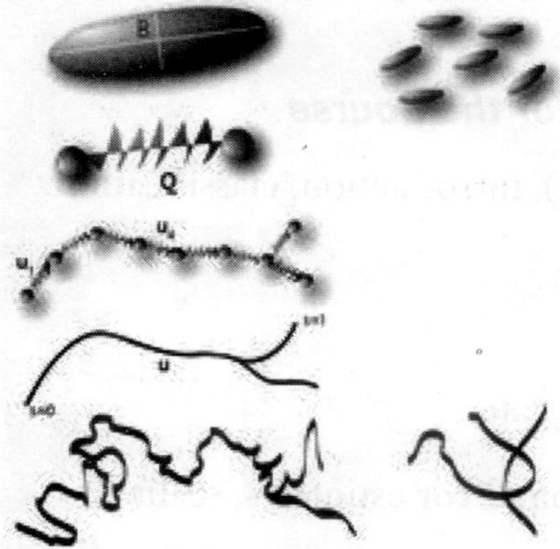

Fig. 1.1: 2008 Spring semester course 327-5102-00L.

Teaching Goals

The course offers an introduction to computer simulation methods and their foundations for the physics and material behavior of simple and complex materials and in particular polymeric and anisotropic liquids. The lecture is devised for students which have attended the course 402-0809-00L

Introduction to Computational Physics. The goal is to i) introduce theoretical approaches used in soft matter physics, ii) motivate coarse-grained models, and ii) the numerical solution of many body problems for optical, mechanical, or electromagnetic applications.

Summary and Outline

- *Theory:* equilibrium statistics, nonequilibrium statistics, tensors, nonequilibrium dynamics of anisotropic fluids, partial differential equations
- *Models:* Phase transitions, dumbbell models, chain models, primitive path models, elongated particle models, connection between different levels of description
- *Simulation:* Monte Carlo, brownian dynamics, molecular dynamics, smoothed particle dynamics, embedded atoms, structure recognition, optimization

The lecture focuses on particle methods. Techniques such as Monte Carlo, equilibrium, beyond-equilibrium and nonequilibrium molecular dynamics, smoothed particle dynamics, dissipative particle dynamics, Brownian dynamics, embedded atoms, lattice Boltzmann will be introduced and applied. Master equations, Markov processes, Fokker-Planck equations, stochastic differential equations play a major role in the fundamental chapters. Substances: from simple towards structured fluids (gases, polymers, ferrofluids, liquid crystals, glasses, metals). The lectures and exercises will be structured as follows:

Tentative Schedule for the Course

Cellular Automata (CA), introduction, classification:

- Moore models
- Traffic modeling
- Shock waves
- The Q2R Ising Model

Monte Carlo integration, error estimates, scaling:

- Monte Carlo in statistical physics, Quasi Monte Carlo
- Distributed random numbers, methods
- Accretion
- The Ballistic Deposition Model
- Diffusion limited aggregation (DLA) model
- Fractal dimensions, polymers

Random walk, theory, applications:

- ❖ Monte Carlo simulation for the random walk
- ❖ Freely rotating polymers
- ❖ Wormlike polymers
- ❖ Self-avoiding walk (SAW)
- ❖ Slithering Snake, Pivot, Enriched samples algorithms

Master equation, introduction, applications:

- ❖ Stationary solution of the Master equation
- ❖ Detailed balance
- ❖ Coupled equations for moments
- ❖ Metropolis Monte Carlo (MC)
- ❖ Thermostatted 1D harmonic oscillator via Metropolis MC
- ❖ Lennard-Jones system via Metropolis MC
- ❖ Gaussian integrals

Ising model and Phase transitions:

- ❖ Metropolis MC algorithm for the 2D Ising model
- ❖ Finite size effects
- ❖ Phase transitions & percolation
- ❖ Percolation theory in the Ising Model
- ❖ Q-Potts model and foams
- ❖ Random site percolation model

Lattice gases, method, simulation and applications:

- ❖ Lattice-Boltzmann, method, simulation and applications

Multiscale dynamics:

- ❖ Molecular dynamics, implementation, applications
- ❖ Mesoscopic interaction potentials
- ❖ Finite difference methods
- ❖ Einstein frequency and configurational temperature
- ❖ Periodic boundary conditions
- ❖ Temperature control
- ❖ Lees-Edwards boundary conditions

Beyond-equilibrium molecular dynamics:

- ❖ Long-range forces
- ❖ White and colored noise
- ❖ Random vector with given mean and covariance

- Whitening a random vector
- Brownian dynamics, theory and simulation
- Langevin equations

Rouse model for polymer solutions:

- Reptation models for polymer melts
- Pompom-type models for polymer melts
- Flow channel, flow birefringence
- Dendronized polymers, simulations, experiments, applications

Kramers process:

- Brownian dynamics of a FENE dumbbell
- Smoothed particle dynamics, soft fluid particles
- Dissipative particle dynamics

Ferrofluids, ferromagnetic chains, theory, simulation, applications:

- Magnetorheological fluids, simulation methods
- Liquid crystals, theoretical approaches, simulation, experiments
- Liquid crystalline polymers, theoretical approaches, simulations, experiments
- Frank-Ericksen elasticity, theories, experiments

Wormlike micelles, living polymers, theory and simulation:

- Plasticity, metals, embedded atoms simulations
- Phoretic forces

Coarse-graining procedures for polymeric systems, shortest paths:

- Delaunay triangulation and Voronoi diagram
- Hyperbolically regularized hydrodynamics.

Script

A script, exercises, sample codes and other supplementary material will be available online.

1.4 Computational and Experimental Mechanics (CEM)

Polymer Technology

The research is aimed at bridging the gap between science and technology in the area of polymer processing and design, through the use of experimental and computational tools in the modeling of the full thermo-mechanical history of material (elements) during their formation, processing and final design, to quantitatively predict properties of processed objects

Computational Rheology and Multi-scale Modelling

Rheology is the science that is occupied with the behavior under flow (deformation) of fluids that cannot be described by a simple linear relation between deformation rate and resulting stress (non-Newtonian fluids). The range of applications is vast and encompasses the spectrum from polymer- and food processing to flow of blood and of DNA through microdevices. Particularly important for industrial processing are polymer fluids such as melts and solutions, which are viscoelastic.

Several clear trends emerged during the last couple of decades. First, use of computational methods, like the finite element method, has become indispensable for applying (model) rheological insights to practical flow problems. Furthermore, the most successful constitutive models in simulations have proven to be those that include microstructural features. This has motivated researchers to try to bridge the gaps between length- and time scales. Ideally, one would like to start at the molecular level (the very architecture of molecules) and end with a simple model that can be used to accurately predict macroscopic flows, and, of course, ultimately the properties of the resulting (solidified) materials obtained. This has not yet been realized. Therefore, our focus is threefold:

1. Development and implementation of numerical methods that are particularly suited for solving the flow of viscoelastic fluids. These methods are usually based on finite element methods. Although our focus is on development and implementation, it is strongly motivated by practical problems from industrial environments. For instance, in polymer processing, various instabilities can occur that limit the flow rate; and in order to predict these instabilities, efficient, accurate and numerically stable schemes need to be developed, that must also be suited for realistic fluid models.
2. Simulations at a mesoscopic level in rheometric type of flows (computational rheometry). For example, by studying suspensions of hard particles in polymer fluids by direct simulations it is possible to obtain information on the average constitutive behavior of such systems. In addition, it enables study of the detailed "down-stream" behavior, such as crystallization and resulting morphologies of solids.
3. Development of computational techniques to bridge the length-scale gaps (multi-scale modeling). It is usually difficult to bridge the gaps by a simple

model. An alternative is to solve for two scales at once. For example, one can start with a macroscopic mesh for the flow and at the integration point level substitute a stress computed from a mesoscopic simulation in a rheometric type flow developed under theme 2.

Applied Rheology and Process Modelling

The need for faster product development and shorter time-to-market, combined with the increasingly higher demands in product and process design, requires a change from experimental trial and error to quantitative predictive capability. The main objective of applied rheology is to provide knowledge and models for the prediction and understanding of final properties of polymeric products. These properties are determined by intrinsic (molecular) material parameters and, to a great extent, by the processing conditions. To reach this goal, applied rheology aims at bridging the gap between sophisticated rheological models and reliable, predictive modeling of polymer processing. Examples of topics in this research area are:

- ❖ Development of non-linear viscoelastic models,
- ❖ Flow-induced crystallization in semi-crystalline polymers,
- ❖ Rheology of two-phase flows,
- ❖ PVT behaviour under high shear and at high cooling rates, and
- ❖ Modeling of injection moulding.

Experimental characterization and validation are important items. For example, reliable elongational viscosity data and data on flow-induced nucleation are not available for most polymers. For validation, specially designed flow cells that mimic the flow in polymer processing operations are used in combination with LDA, PIV, FIB measurements and real-time X-ray and Raman spectroscopy. In this way the influence of molecular parameters, such as molar mass distribution and molecular architecture, on flow, nucleation and crystallization behaviour of polymer melts can be evaluated.

Chaotic Mixing and Two-Phase Flows

Deformation, coalescence and break-up of fluid volumes, phase separation, phase inversion and interfacial deformation solely, are important physical phenomena that determine the mechanical and optical characteristics of multi-component flow and composition. In case this flow results from an industrial process, the process conditions required and the product properties obtained, will strongly depend on these flow and composition characteristics. In addition to relevant experimental investigation, suitable physical models, together with an appropriate implementation in a numerical code, are important to obtain better understanding of the physical phenomena mentioned above. In particular we develop mapping methods and extend diffuse interface modeling.

1. The mapping method (MM) for single-component fluid systems is designed to deal with the optimisation of distributive mixing flows. The model is applied to spatial- or time-periodic flows where the kinematics of the mixing process is written as a mapping between two reference grids. Classical (particle tracking) and non-classical (parallel processing) techniques are combined to construct matrices, which represent this mapping. The method is general and is successfully applied to optimise flows ranging from simple two-dimensional time-periodic prototype mixing flows to geometrically complex three-dimensional mixers.
2. The diffuse-interface model (DIM) for multi-component fluid systems is derived from an elegant theory, which couples thermodynamics with hydrodynamics and results in two extra variables, the chemical potential and the composition, determined by two extra equations (a convection diffusion equation for the composition and an equation that relates the local chemical potential with the local composition of the fluid). One of the fascinating features of this model is that it combines the physical phenomena described above and, for example, is applied to study the blending of phase separating polymer systems during flow.

Applications of this work are found in chaotic mixing in micorfluidics, self-stratifying coatings (phase-separation at interfaces), inkjet design for fast, accurate printing or copying, processing of organic super-fibres that outperform steel in specific strength (and modulus) by one order of magnitude, food processing, polymer blending and processing, multi-layered structures etc.

Structure-Property and Constitutive Modeling

The research focuses on the development of new polymeric systems with optimal mechanical properties (toughness/strength/durability). To achieve this goal it is required that modeling of the macroscopic response of the polymeric material is based on its molecular and microscopic structure, as on those length scales materials can be controlled. This length-scale bridging is achieved by :

- ❖ Use of new, experimentally validated constitutive equations that capture the important features of the polymers' intrinsic deformation behaviour, and
- ❖ Homogenisation techniques based on representative volume elements with periodic boundary conditions (multi-level FEM) that are employed to analyse the macroscopic mechanical response.

The three-dimensional non-linear viscoelastic constitutive equations are derived and improved on the basis of experimental observations and include yield, intrinsic strain softening, strain hardening and the important dependence on strain rate and temperature. New developments include the assessment of model parameters on a microscale using indentation techniques, and a numerical tool that can effectively predict the development of mechanical properties during processing of glassy polymers. Subsequently, the macroscopic response of heterogeneous polymeric systems is predicted, based on the micro-deformation. In determining the intrinsic properties of polymeric materials, a distinct length scale is experienced that shows that thin films behave different as compared to the bulk. Given the importance of thin films in modern electronic devices, e.g. poly-LEDs (light emitting diodes), LCD's (liquid crystal displays) and sclar cells are based on thin films, special attention is paid to their characterization and constitutive modeling. New routes are explored to obtain specific structures of commonly brittle polymers that display a dramatic increase in their strain to break.

- ❖ Structure-Property Relations and Constitutive Modelling of Amorphous Polymers.
- ❖ Structure-Property Relations and Constitutive Modelling of Semi-Crystalline Polymers.
- ❖ Multi-scale mechanics of Polymers.
- ❖ Micro-Wear of Polymers.

Surface Mechanics and Microfluidics

Thin layers, surfaces and coatings are of great importance in different kinds of engineering applications. Nonetheless, the principles of friction and wear are still poorly understood. Analogously, the properties of thin layers and surfaces are different from those in the bulk materials, and these differences are equally ill-understood. In this new starting topic we attempt to better characterize and understand these properties for polymers and metals, and combinations thereof. The methods employed in this research are optical microscopy, environmental scanning electron microscopy (ESEM and high-resolution SEM), orientation imaging microscopy (OIM) in combination with in-situ loading, nano-indentation techniques, lateral force measurements, single-asperity experiments on flat and well-defined surfaces, JKR techniques to study the origins of adhesion and, finally, molecular modeling techniques.Similarly, in microfluidics the analyses are different from those of the bulk and sometimes we can make use of these scale effects. We study the scale effects in solid mechanics, in close cooperation with theme 4, and of fluid mechanics, in close cooperation with theme 3.

- ❖ Nano-indentation of Polymers
- ❖ Micro-Tribology of Polymers
- ❖ Chaotic Mixing in Micro-Channels
- ❖ Active Surfaces in Micro-Fluidics

Mission

The mission of the Materials Technology group is to bridge the gap between science and technology in the area of materials processing and design, via the use of computational tools in the modelling of the full thermo-mechanical history of material (elements) during their formation, processing and final design, in order to be able to quantitatively predict product properties.

About CEM

Within the program Computational and Experimental Mechanics (CEM) we apply principles from fluid and solid mechanics to a diversity of problems in the design of materials and precision products made of polymers and metals and combinations thereof, including the modeling of the manufacturing processes necessary to make the products. Multi-scale solid mechanics plays a crucial role, since it couples structure to properties. Fluid mechanics is important, since the final structure is generally obtained by mixing components in their molten state or by using liquid-liquid or solid-liquid phase separation. On the other hand it is also obvious that the processing determines the resulting structure, which renders rheology in complex geometry's a key-area.

Two Issues Prove to be Crucial

1. First, the different relevant length scales involved require the solution of how to bridge those scales from the atomic or molecular, via the micro-structural to the macroscopic level and vice versa. Special multi-scale numerical techniques are being developed to solve this problem. A multitude of microscopes, from light microscopy, via environmental scanning electron microscopy to atomic force microscopy, is used to study in-situ deformation on the different length scales involved. Besides, at the ESRF, the european synchroton radiation facilities, in Grenoble we study with time-resolved experiments on the micro-beam line changes during deformation with X-ray scattering techniques.
2. Second, the complete thermal-mechanical history experienced by the material during processing and shaping, clearly influences the final

mechanical (and e.g. optical) product properties. Advanced modeling thereof is required to obtain quantitative predictions and provide tools for optimization of process and product design. With polymers, structure development during flow in complex geometry's is the key element to be revealed and different experimental (LDA, PIV, PTV, ROA, etc.) and numerical (FEM, BEM, SEM, etc) techniques are used, mostly in industrial prototype flows. Analogously do the mechanical properties of products in metal engineering largely depend on the processing history, where grain size, dislocation structures, phase transformations and complex interactions between the heterogeneous phases play a crucial role. Mixed numerical-experimental analysis of the evolving microstructures is a key factor in establishing structure-property relations. They basically can be considered as the future design tools for manufacturing processes and product optimization in high technology applications.

Research within this program is performed in the technological top institutes DPI (Dutch Polymer Institute) and NIMR (Netherlands Institute of Metal Research) in cooperation with industry and with similar university groups in Europe and the USA.

2

Theoretical Polymer Science and Technology

2.1 Classical Polymer Processing

Subsequent to half a century of productive and intensive development, the field of *polymer processing* has reached a turning point. Contrary to the initial expectations of the gatekeepers of this field of engineering activity, polymer processing didn't turn into another well-defined engineering discipline, with a formalizable and teachable body of knowledge, alongside chemical, mechanical, civil, electrical, and other more or less classical engineering disciplines, as it happened, for example to biomedical engineering and computer engineering. Indeed there are very few universities, if any, that have a Department of Polymer Processing, which grants undergraduate level academic degrees in that field. Rather, polymer processing is mostly taught at the graduate level predominantly to students with chemical engineering and mechanical engineering background.

This development didn't happen because discipline lacked the key components of an 'engineering discipline': a unique body of knowledge; the practitioners working in a well-defined segment of the industry; professional societies promoting the discipline; dedicated scientific journals disseminating knowledge; a global college of researchers and practitioners; and a unique character, texture and jargon.

This development did not happen because of a shrinking marketplace making the discipline less relevant or obsolete. The opposite is true. Polymer production continues to grow and even in the USA, which has the highest per capita consumption of polymers in the world, saturation level is expected to be reached only in the middle of the 21st century, as shown in Figure 2.1. And, since every single pound of polymer manufactured must go through one or more processing machines, growing production implies a growing polymer processing industry.

Not only does the production of plastics in absolute numbers continue to grow in the USA, but also the consumption of plastics when normalized by the GDP continues to grow at an impressive rate. In fact plastics is the only material, which continues to grow when normalized by the GDP, as shown in Figure 2.2.

Total Polymer Production in the USA 1934-1991 (ton/y)

Saturation Level =84,500
Process midpoint 1981
Logistic Projection

100000
80000
50000
40000
2000
0

1930 1940 1950 1960 1970 1980 1990 2000 2010

Fig. 2.1: The production of polymers in the USA in kilo tons per year, based on research carried out at the Rockefeller University in New York.

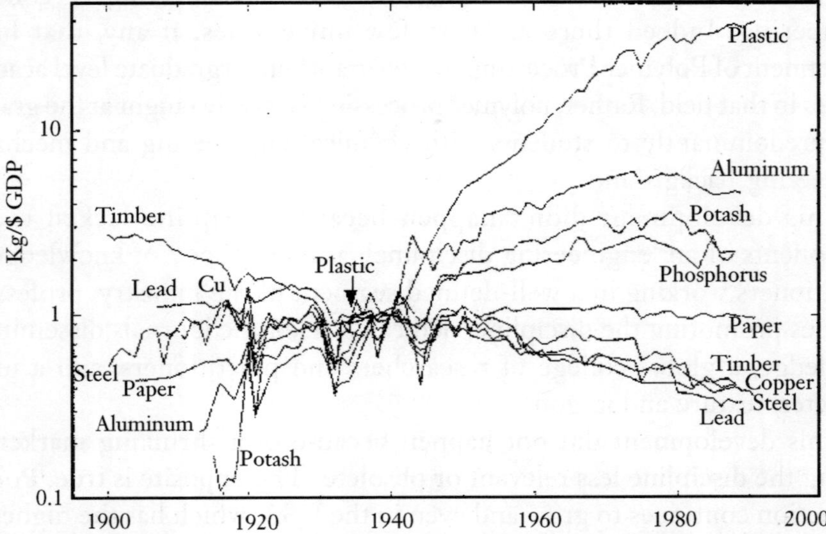

Fig. 2.2: The consumption of materials normalized by the GDP, based on research carried out at the Rockefeller University in New York.

The current total world consumption of polymers is estimated to be 200,000 kton per year, which is approximately 2.5 times the USA consumption. There are no comparable estimates on world-wide consumption saturation levels, but considering the much lower per capita consumption in the less developed world, the much lower GDP per capita as shown in Figure 2.3, and the still quickly growing human population, which is expected to reach 10-12 billion people by the end of the 21st century, and considering the fact that today only about 5 per cent of the oil is used for polymer manufacturing leaving huge reserves for increased production of plastics, it would be reasonable to assume that world-wide plastics production will continue to grow at very significant rates throughout this century.

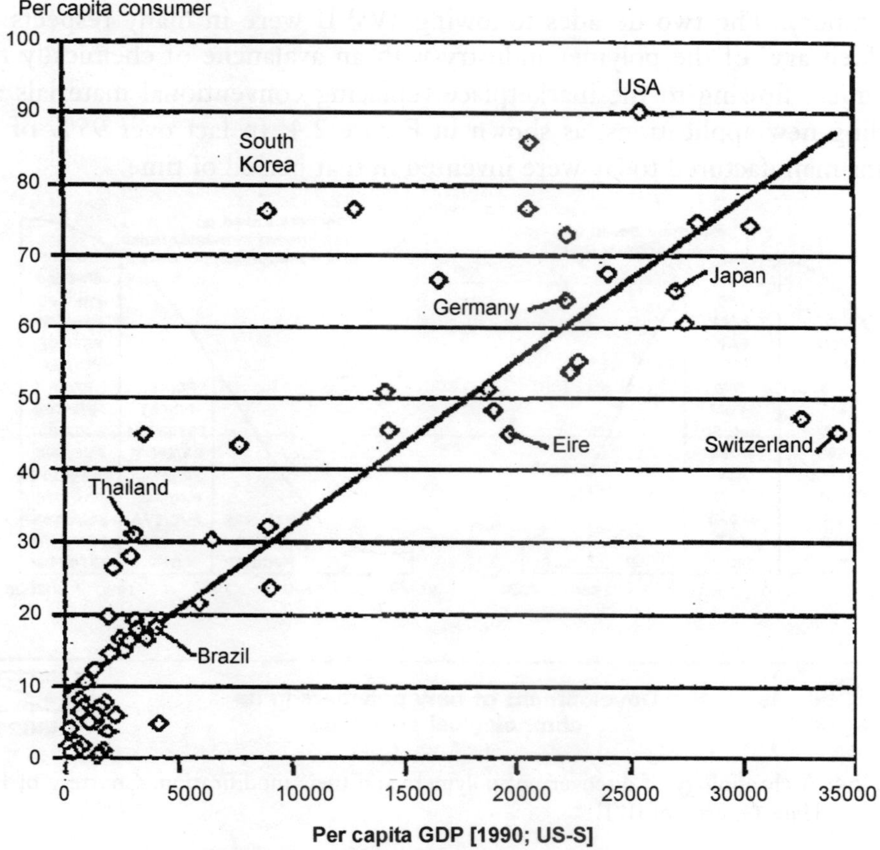

Fig. 2.3: Polymer per capita consumption versus GDP per capita from H.U.Schenk in "Industrial Polymer Research" AIM Magazine July 2001.

Why then didn't polymer processing gel into a distinct engineering discipline as it occurred to other engineering disciplines that emerged in the course of the past century and half?

To answer this question we must take a historical perspective and review in a nutshell the developments in this field over the past half-century, and briefly also review some mega trends in the chemical and polymer manufacturing industry.

The Golden Age

Polymer Production

World War II, with the discovery of low-density polyethylene at ICI (which was a crucial component for the development of radar) and nylon at DuPont, marks the beginning of the modern polymer industry, though, of course, PVC, PS, celluloid, and thermosets such as Bakelite preceded it, as did the rubber industry. Indeed most of the polymer processing machinery derived from rubber machinery. The two decades following WWII were in many respects the "golden age' of the polymer industry with an avalanche of chemically new polymers flowing to the marketplace replacing conventional materials and finding new applications, as shown in Figure 2.4. In fact over 95% of the resins manufactured today were invented in that period of time.

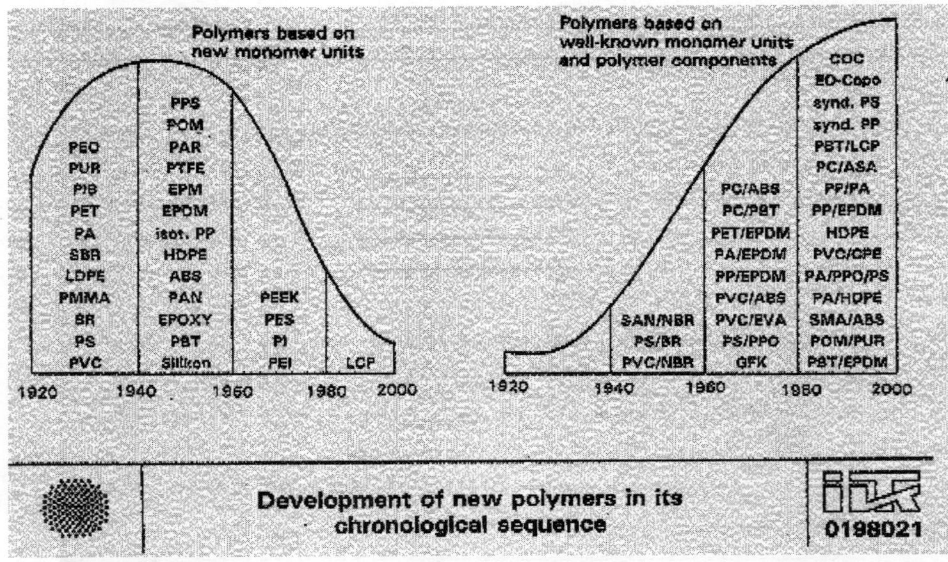

Fig. 2.4: A chronology of discovery of polymers and their modification. Courtesy of Prof. Hans G. Fritz of IKT.

In this period some 40 chemically new polymers were developed. Polymer chemists claim that this is an enormous arsenal if we could learn from nature how to manipulate them into appropriate conformation and spatial arrangements. Just consider, they say, the vast variability that nature came up with only some 20 amino acid building blocks!

Polymer Processing Education

This was also the period of the formal theoretical formulation of the discipline. In the United States, the first formal academic formulation of polymer processing as a discipline came with E.C.Bernhardt's collective contribution of DuPont scientists in the 1959 *Processing of Thermoplastic Materials*. Later, in 1962, Jim McKelvey took the discipline one step further and incorporated a more fundamental approach of the transport phenomena borrowed from chemical engineering. Both, however, retained a chemical engineering "unit operation" like approach by treating extrusion, injection molding, calendaring etc. as identifiable and separate "unit operations", which make up polymer processing. Tadmor and Gogos in the 1970s departed from this notion and developed an engineering science-based but uniquely polymer-processing rooted approach by introducing the "building blocks" of polymer processing, namely, the "elementary step" which are typical mechanisms common to all machines and processes, the "forming steps" and "structuring step". There were of course many other books, some general and some specific, but they don't alter the overall picture.

Polymer Processing Research

The focus of polymer processing research during that period was on trying to elucidate, understand and mathematically model mechanisms taking place inside processing equipment such as single and twin-screw extruders, injection molding machines, calendars, continuous mixers, blow molding machines etc. utilizing the well-established *Transport Phenomena* approach of Bird, Stewart and Lightfoot. Such mechanisms determine the rate and efficiency of the operations such as melting, conveying, mixing, and devolatilization; non-isothermal flow of molten polymers through dies and into closed spaces such as mold filling. For doing this with accuracy sufficient for machine design there wasn't a need for the very sophisticated viscoelastic rheological models and certainly not for molecular theories. These, however, were needed for modeling free flowing steps such as fiber spinning, film blowing and blow molding.

While modest rheological, and polymer physics and chemistry tools were sufficient to describe *quantitatively* most of the phenomena occurring during the "elementary" and some of the "forming " steps, through simulation-solved "design algorithms" they were hardly adequate to describe quantitatively the "structuring" steps, which are responsible for the processed product properties, short and long-term: they involve an intimate knowledge of macromolecular conformation and relaxation from initially non-uniformly oriented melts and retained stress-induced polycrystalline morphologies nucleation and crystallization rates in non-uniform temperature fields following die flows and mold filling. These require advanced molecular-based constitutive relations, and quantitative understanding/description, of crystallization phenomena under dynamically changing molecular orientation fields. Thus, the goal of being able to model and predict the effect of "structuring" on final past properties remained the "Holy Grail" in that period.

During these decades, therefore, it appeared that the field of polymer processing is gradually expanding its scope by creating more and more knowledge specific and inherent to its character and nature and absorbing somewhat less from other disciplines, thus sharpening its unique identity and evolving-converging into a well-defined and even a reasonably mature engineering discipline. Indeed, the latter part of this period, the closing decades of the 20th century, was also characterized to some extent, at least from a theoretical point of view, by stagnation not uncharacteristic of other mature engineering disciplines.

Parallel Developments in Polymer Rheology and Polymer Sciences

Throughout this period and beyond, the relevant sciences and engineering sciences such as rheology, polymer physics, polymer chemistry and computational techniques, feeding the polymer processing discipline have undergone explosive developments. Thus, rheology, which is part of physics, encompassing continuum theories and molecular theories, characterization and measurements, grew by leaps and bounds. It was carried out mostly by scientists who engaged in it because of scientific curiosity, and who by-and-large had little interest in polymers or polymer processing as such, but who appeared to be pleased that the polymer and the processing industries existed, as they could justify (to funding agencies) that their work is significant for the industry in the processing of polymers. And, it was. But, for a quarter of century the gap between the science rheologists developed and the elements of science that could be applied to polymer processing with the tools available at the time remained very wide. Similar developments took place in polymer physics and chemistry and instrumentation. This growing body of knowledge is bound to play an important role in future developments of the field.

The Golden Age of polymers was an immensely fruitful period. It laid the foundation and the knowledge base for a new discipline which appeared step-by-step to formulate its own "bag-of-tools", character and scientific foundation, jargon and international college of practitioners and researchers, professional societies etc.; namely all the essential components to become a recognized engineering discipline. But, this didn't happen. However, before analyzing the reason why not, some developments in the larger chemical industry and resin manufacture industry ought to be very briefly reviewed.

The Restructuring of the Chemical and Resin Producing Industry

The global chemical industry is vast and amounts to $1.7T sales, contributing about 4% to global GDP. For decades the chemical industry in the US and elsewhere was one of the most innovative industries, and chemical engineers commanded the highest average salaries. But, maturity and globalization changed this picture. Market driven stiff global competition and ongoing demand for increased shareholders value forced the chemical industry and the resin producing industry into a massive restructuring process. Cost cutting of all types (including corporate R&D), benchmarking production costs, building large shares in relatively few products, globalization of the business, and moving whenever practical manufacturing close to the source materials, were among the steps taken. These indeed improved profitably but were quickly exhausted, and it became recognized that the lack of innovation was the main source of the shrinking profit margin. Consequently there began a massive move toward life sciences and specialized products.

However, during the same period of time, over the foregoing backdrop of a restructured resin manufacturing industry, we also began to witness a number of major developments that appear to profoundly change the overall picture bringing about the 'turning point'.

The Turning Point

The first was the development in the 1970s and 1980s of the novel homogeneous single-site (metallocene-based) catalyst for polyolefins, and further developments of catalytic polymerization of polyolefin based polymers and copolymers. These gave enhanced control over the molecular structure and brought a resurgence of the polyolefin business, which is no longer just a commodity polymer production business.

The second development at about the same time was the progress in high performance engineering plastics, specialty polymers and the emergence of functional polymers such as electric conductive and light emitting polymers upon application of voltage, photo addressable polymers, etc.

The third development is an on-going search for new, diverse and better properties by what could perhaps be best defined as *scientific compounding* by creating polymer blends and alloys, copolymer, block-copolymers graft copolymers etc. via reactive processing, by incorporating special solid, liquid and gaseous additives, and by initial attempts to control the conformation and macrostructure of the polymers as shown in Figure 2.4. The uniqueness of this development is that most of it is carried out within processing machines creating either "designer pellets"—as coined by Costas Gogos, namely compounded pellets which carry the desired properties to the finished product processed in the final forming step, or as the trend seems to be, to fuse the compounding with the forming, i.e., in-line compounding step.

The fourth development is related to the computer revolution placing enormous computational power at affordable cost at the disposal of the researcher and the practicing engineer. This led to quick development in finite element computational and representational techniques, making computational fluid mechanics a hot thriving field. Consequently, finite element formulation in complex geometries with complex rheological constitutive equations under transient and non-isothermal conditions, as experienced in polymer processing, became practically possible. Moreover, this new capability also enabled in principle, to follow the thermo-mechanical history of fluid particles and predict their properties all the way down to molecular level. Indeed the mission statement of the Eindhoven Polymer Laboratories and Material Technology Center headed by Professor Han Meijer: "*to bridge the gap between science and technology in material processing using modeling and computation of the full thermo-mechanical history during formation to quantitatively predict properties*" elegantly captures these new trends.

The vision statement of Clemson University Center for Advanced Engineering Fibers and Films (CDEFF) headed by Prof. Dan Edie echoes that of Eindhoven CAEFF promotes the transformation from trial-and-error development to computer-based design of fibers and films. This new paradigm for materials design—using predictive numerical and visual models that comprise both molecular and continuum detail—will revolutionize fiber and film development.

The fifth development has to do with the explosive developments and growing sophistication of sensor instrumentation for quantitatively measuring rheological properties, during flow and deformation and physical and mechanical properties down to the molecular level, and their chemical composition.

The foregoing changes redefine the field of polymers and place new demand and new challenges on the processing and formulation of polymeric materials. Not only understanding what happens to a given polymer in the machine with the purpose of designing better and more efficient machines as was the case in the 1960s and 1970s, but also manipulating the molecular structure and conformation of the polymeric materials, their chemical composition or altogether creating modified new materials in the machine. So polymer processing, as a discipline is no longer concerned with just "*conversion of the raw material into a useful product*", in the narrow sense of the world, but by using all the arsenal of modern science to create new and better materials in the process. This finally fuses polymer processing with polymer physics and polymer chemistry, with the powerful new computational tools providing the glue to make this fusion qualitative, modellable and predictive, and with the modern instrumentation experimentally verifiable. *Hence polymer processing, rather than converging into a new discipline, diverged into a broad multidicplinary activity much like biotechnology did.* This multidisciplinary field can be best described as *macromolecular technology or macromolecular process science*, which includes the processing of all the material that have a macromolecular structure, both natural biological macromolecules and synthetic macromolecules.

In fact what appears to have happened is that *before* polymer processing had a chance to converge into a well-defined engineering discipline, it diverged into a multidisciplinary activity as it happens to other leading engineering disciplines. This is the inevitable result of the exponential growth of science and the merging of science and technology into an indistinguishable new entity at the closing decades of the 20th century as discussed below.

The foregoing mission statement of the Eindhoven laboratory, CAEFF, as well as that of CEMEF headed by J.F.Aggasant and possibly others capture this broad new multi- disciplinary nature of *macromolecular technology*, and also point toward a major development that is needed, but didn't happen as yet, and this is, the scientific and quantitative closing of molecular composition and structure and final properties. This should become a fruitful research area, as it has in metal processing.

This outward divergence and multidisciplinary nature of the processing of macromolecules has, of course, profound implications on research and development and the future of the industry.

Multidisciplinarity

The fact that polymer processing did not converge into a discipline, but rather diverged into a powerfully fruitful multidisciplinary activity, perhaps should not be too much of surprise because this is the dominant trend in many frontier fields in science and technology. It would be, therefore, worthwhile to briefly examine the reasons for this trend and the nature of multidisciplinarity.

Knowledge in science is one, but human limitation led to its subdivision into well-defined manageable segments or disciplines. This was an inevitable and profoundly important step in the evolution of science, because it eliminated dilettante scientific research and enabled in-depth exploration of nature by professional scientists resulting in an enormous growth of knowledge. Each discipline in science developed its own specialized approach, jargon, and interconnected global network, which remain essential elements in evaluating, maintaining and promoting scientific excellence.

In engineering a similar development took place. Much of the faculty research in engineering at the leading universities, throughout most of the 20th century, focused on analyzing existing technologies with the powerful tools of science and defining the scientific principles of the engineering disciplines. These principles embodied in the *engineering sciences* form the *core* of current engineering education and practice.

However, as we approach the end of the 20th century, disciplinary research, focused on elucidating and mathematically formulating elements of the particular engineering science itself rather then the real technological problem 'out-there', in most fields of engineering has become somewhat less productive, and the need for multidisciplinary research for tackling real and increasingly complex problems is becoming mandatory. In the sciences, a similar development is taking place.

The preoccupation with disciplinary research reflected a natural stage in the development of engineering and science. This stage catalyzed by WWII by the generous defense-driven funding rationale formulated by the famous Vannevar Bush report entitled "Science the Endless Frontier" and driven by superpower defense oriented policies, was enormously productive in the past half century. However, with the collapse of the Soviet Union and the end of the Cold War, economic competitiveness replaced this funding rationale. Over the backdrop of an on-going nearly total fusion of science and technology in the last quarter of this century, this new economic-driven rationale also inevitably called for solving more complex science-technology problems, which in turn mandate multidisciplinary research. In addition the fusion of science and technology also lead to the abandonment of the 'linear model', and blurred the distinction between 'basic' and applied' research.

Researchers in most disciplines, therefore, are beginning to recognize that their own disciplinary "bag of tricks" may not be sufficient to advance new frontiers. New advanced research in technology-science, by its very nature, increasingly requires a multiple of disciplines to be explored. This cannot be accomplished by individual researchers superficially familiar with the relevant disciplines, but rather by collaboration of groups of researchers, each an expert in his/her relevant discipline. For example, how could optoelectronics be studied without close interaction between physicists, chemists and electrical engineers; high temperature superconductivity without cooperation between physicists, chemists, chemical engineers and materials scientists; *nanotechnology* without the collaboration of scientists in solid mechanics, physics, computer engineering, electronics, and biochemists; or the search for biologically based computing devices be conducted without the mutual support of molecular biologists, physicists, biochemists and computer engineers? Moreover, there are engineering and scientific fields, which are multidisciplinary by their very nature. Materials science and biotechnology are cases in point. And, now polymer technology joins this rank.

Finally, the increasing university-industry collaboration, which meets both the economic-driven rationale and the societal rationale implying increasing consideration to societal needs, tends to require multidisciplinary research as well. These developments have lead to a new academia-government-industry alliance, which is the only one that can secure the massive funding needed for carrying out cutting-edge multidisciplinary research. Thus, perhaps, ushering in a new phase in the development of universities with the Government and Industry jointly acting as "patrons" of the universities, replacing the Government as the sole "patron" as it has been the case for a considerable time, which in turn historically replaced the Church as the sole patron of universities.

If multidisciplinary research on the university scene is becoming the common mode of research rather than the exception, it is worthwhile to explore the issues this raises in current university structure and culture.

The assimilation of multidisciplinary research into the university fabric requires urgent and important changes. To begin with, it requires a new institutional culture encouraging the individual faculty member to cross disciplinary boundaries, overcoming the current prevailing academic culture of isolated academic departments that behave like tiny "nation states". Special attention should be focused on young faculty members, who are more inclined for multidisciplinary research, but are hesitant to do so because possible negative implications on their promotion. The individual faculty member should, on one hand, remain firmly rooted in his or her discipline, but at the same time, he or she should routinely cross disciplinary boundaries in both research and teaching. Faculty should perceive the departmental unit as a "home base" rather than a "cocoon".

Moreover, the classical "faculty member-graduate student" team and research mode, should be increasingly replaced by a larger team model including "a group of faculty members and a group of graduate students". This model will better prepare the future doctoral student not only for an academic career, but also for an industrial career, as future high-tech industry will require increasing number of Ph.D. level engineers and scientist.

Multidisciplinarity implies a meeting ground of minds with different skills, training and thinking to combine knowledge, talents, and experience for the benefit of all. It is a profoundly democratic concept. Socrates among others practiced it at the Greek Agora long before modern science came to the forefront of human civilization and reshaped the world.

Industry Academia Changing Roles

During the post WWII period, prior to the foregoing changes in the research scene, university research was mostly "basic" research. The major companies who had their own large in-house reaserch laboratories carried out the "applied research" and the development work to convert ideas into products. Their senior researches attended the scientific conferences and maintained close touch with academic research and were the transfer agents of the new discoveries to practice. But, in the new changing environment and in particular in view of the increasing global competition, the major corporation have reduced or completely eliminated their own research laboratories, and reduced their research effort by-and-large to product support and just-in-time research.

These major shifts in orientation, with the new funding rationale have driven universities to do more of what is referred to as "applied research". However, this new unfolding scene matched perfectly inherent development in science and technology. Because the fusion of science and technology into a new entity briefly mentioned above, anyhow blurred the distinction between basic and applied research. Term like "pure technological research", "pre-competitive generic research", "strategic research" etc., which are multidisciplinary in nature, capture the spirit of the new research scene on university campuses. Moreover, the lack of in-house "long-term" research in the large companies, yet their need for such research, forces them to sponsor such research at universities and to enter into consortia type collaborative efforts creating a new alliance between government, academia and industry.

The Workshop

The subject matter of this Workshop is to identify and verify the foregoing revolutionary developments that are taking place and discuss their far-reaching implications. To redefine the scope and character of this field in view of these development. To formulate the national, university and industry policy questions that must be asked regarding its future: what are the most promising scientific-technological frontiers for this field; what should the organization and funding of research and development be; what should the nature of the government-academia-industry relations be to provide suitable support; how should scientific publications be modified to catalyze the new developments; what should the nature of international scientific cooperation and interaction be. Hopefully this Workshop will offer some answers and conclusions to these and many other questions to be raised, which will help clearly identify the exact nature of this historical turning point.

2.2 Polyelectrolyte Protein Complexes (NCP)

The nucleosome core particle (NCP) is the elementary unit of the chromatin and plays an essential role in the packaging of DNA in the chromosomes. It is able to condense a Eukaryotic genome of about 2 m length into a cell nucleus, which is only 5 micrometers in diameter. Many cellular functions, such as transcription and replication are effected by the structure of the chromatin, and so understanding the structure of the nucleosome is key to understanding the activation and repression of these functions.

By analytical theory, we study the influence of electrostatic interactions on DNA complexation. As a result, we find that partial charge neutralization of the DNA part touching the nucleosome core leads to spontaneous DNA bending and NCP overcharging.

Dynamics of Biological Macromolecules

Fluorescence correlation spectroscopy (FCS) is an experimental technique particular suited to study the dynamics of biological molecules. Combined with an appropriate theoretical model for the dynamics of the semiflexible macromolecules, this technique provides insight into the dynamics of, e.g., DNA or actin filaments on a segmental level.

The comparison of the theoretical approach to FCS correlation functions of DNA molecules verifies the theoretical model and has allowed us to determine the center-of-mass diffusion coefficients and the longest relaxation times for various DNA molecular weights.

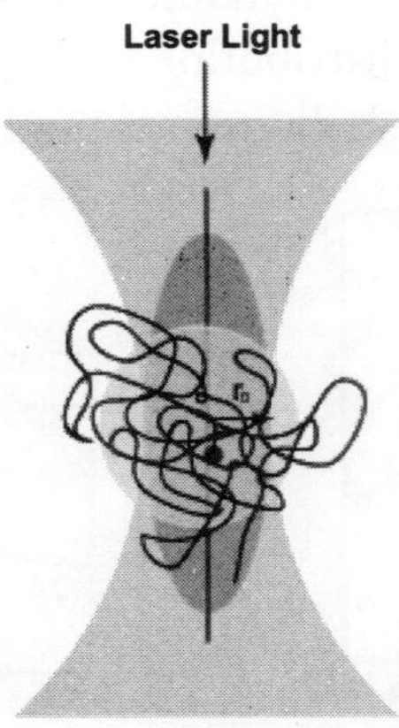

Polyelectrolyte Electrophoresis

Electrophoresis is a standard tool to separate polyelectrolytes, e.g., DNA or proteins, according to their molecular weight. The complex interplay of the various present interactions such as Coulomb interaction, hydrodynamic interactions, chain entropy, external electric field, etc. renders electrophoresis unique among other external fields and explains why this phenomena is far from being understood on a microscopic level.

Mesoscale computer simulations (MPC & MD) provide insight into the properties of charged polymers in solution and reveal the importance of, e.g., counterion condensation and hydrodynamic interactions on polyelectrolyte mobility.

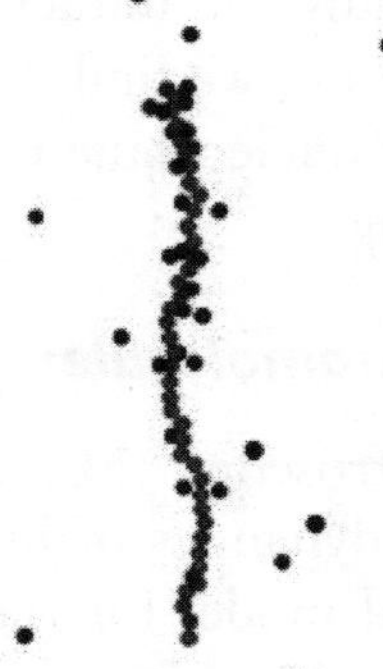

Semiflexible Polymers in Shear Flow

Experimental studies of individual DNA molecules in steady shear flow reveal remarkably large conformational changes and a complex dynamics such as tumbling motion, i.e., a polymer stretches and recoils in the coarse of time. The theoretical analysis of the dynamics of semiflexible polymers by an analytical approach yields the distribution of the orientation angles of the end-to-end vector as well as the distribution of tumbling times.

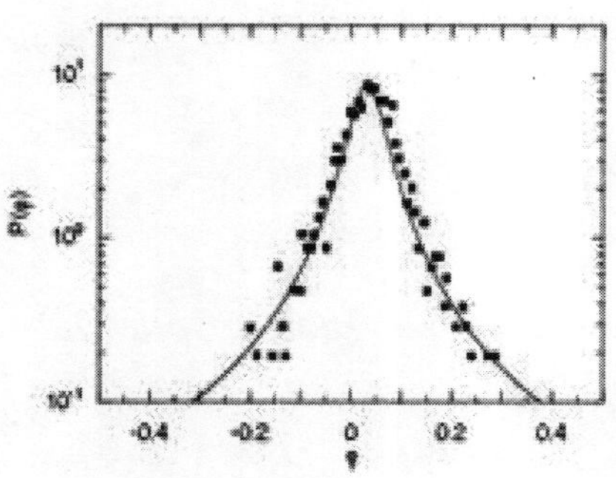

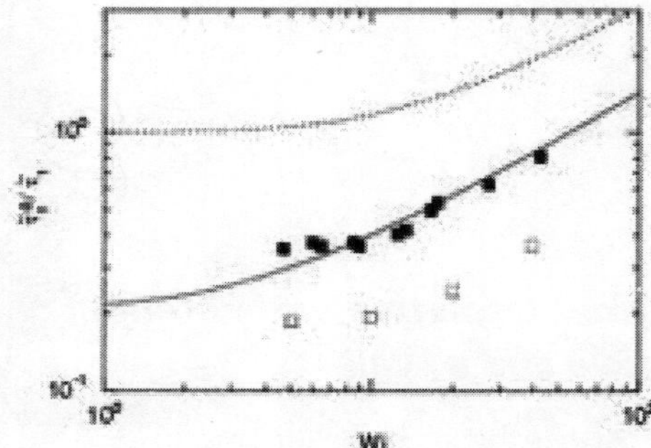

Star Polymers in Shear Flow

Polymers and polymer assemblies exhibit a unique behavior in flow which is related to their conformational degrees of freedom. Their flexibility leads to a simultaneous deformation of the polymer and the fluid flow field, which strongly affect each other. Technologically, these systems are interesting for a variety of applications, such as drag reduction by polymer additives, drug delivery systems, or as motor oil viscosity modifiers. Star polymers, where f linear polymers are anchored to a common center, are interesting because their properties can be tuned by varying the functionality f as well as the arm length.

Using a novel mesoscopic simulation technique, known as multi-particle collision dynamics (MPC), we study the effect of a shear flow applied to star polymers of different functionalities and arm lengths. We investigate the induced anisotropy, orientation and rotation of the star polymers as a function of the flow intensity, as well as the effect of the polymer motion on the surrounding fluid motion.

Small Spherical Particles in a Polymer Solution

The depletion interaction between colloidal particles and nonadsorbing polymer chains has mainly been studied for large particle size and is less understood in the so-called protein-limit of small particle size. At a first glance this interesting case looks complicated since it is dominated by configurations with the chain coiling around the particle and approximations treating the chain as nondeformable fail completely. However, on relating it via the polymer-magnet analogy to a field theory at the critical point, exact results can be achieved for the monomer-density distribution around one and two small particles.

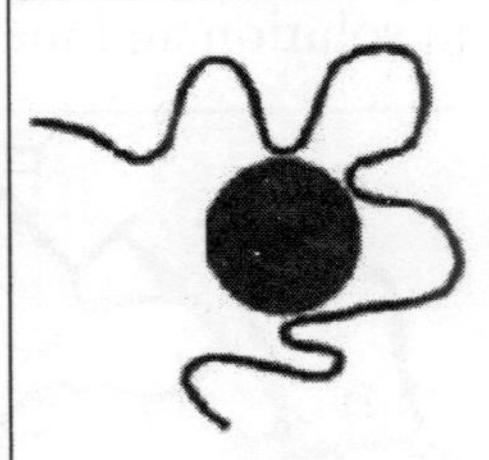

Polymer Depletion and Potential of Mean Force between a Colloidal Particle and a Wall

One of the basic interactions in colloid physics is the force induced between colloidal particles or a particle and a wall by adding nonadsorbing free polymer chains. For entropic reasons the chains avoid the space between the particle and the wall leading to an unbalanced pressure which pushes the particle towards the wall. We find that the particle-wall interaction crucially depends on the particle-to-polymer size ratio p: While for large p the force decreases monotonically with increasing distance between particle and wall, for small p the force displays a maximum.

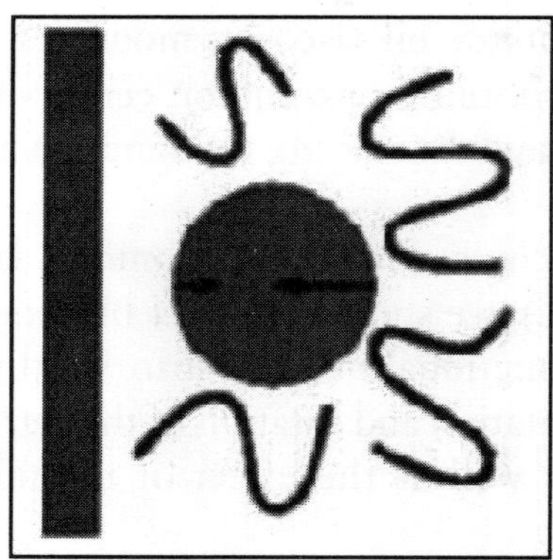

Intramolecular dynamics of linear biological macromolecules (DNA, Proteins, ...).

DNA molecules are rather long polymer chains, thus, the conformations and intramolecular dynamics of individual molecules can be studied experimentally by optical methods. We developed and applied a semiflexible chain model to describe the experimentally observed properties. As a result, we demonstrated that quantitative agreement for the intramolecular relaxation times of a partially stretched DNA molecule can only be obtained, if the force-extension relation of a semiflexible chain model is used, despite the fact that the considered molecule is long and flexible. Quantitative agreement between experimental results and our analytical approach has also been achieved for other properties of molecules in solution and melt.

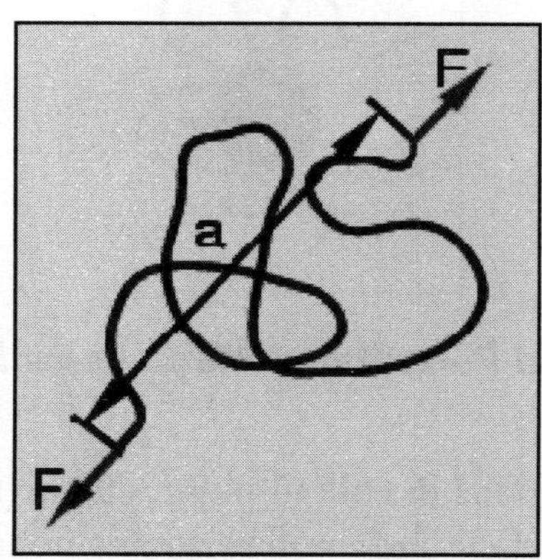

Structure of Polyelectrolyte Systems

The competing interactions among the various components of a polyelectrolyt system determine the structure of the solution and the conformations of a macromolecule. We investigate the structural properties of such systems by molecular dynamics simulations and a liquid state theory (PRISM). From the Ornstein-Zernike relation we obtain the various partial pair correlation functions and structure factors. The calculated effective interaction potential among the polyelectrolyte rods exhibits excellent agreement with the Debye-Hückel potential for small and medium interaction strengths. At large interaction strength, we find an short range attractive interaction between the rods, which is transmitted by the counterions in the system.

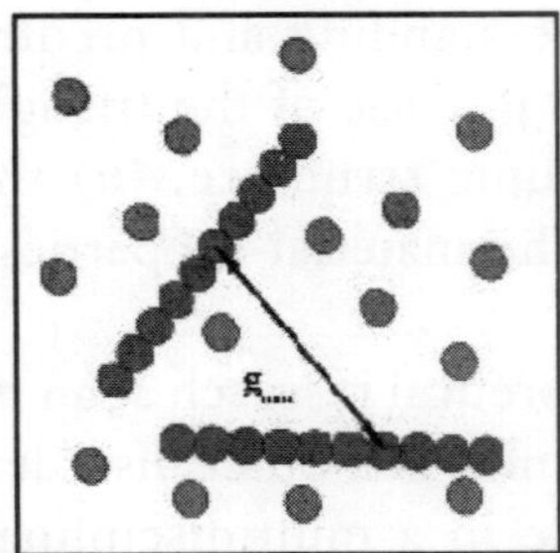

Reptation Dynamics in Polymer Melts

The dynamics of polymer melts and concentrated solutions can be described by the reptation model of Edwards, de Gennes and Doi. We have developed a lattice gas model for reptation which incorporates the effect of constraint release of the entanglement network on the dynamics of a single polymer chain. From this model one can calculate the tube length relaxation of an initally stretched polymer and its fluctuations. The relaxation function is in very good agreement with experimental data obtained by Perkins *et al.* using flourescence microscopy. The lattice gas results for fluctuations indicate that the treatment of noise within standard reptation theory has serious flaws.

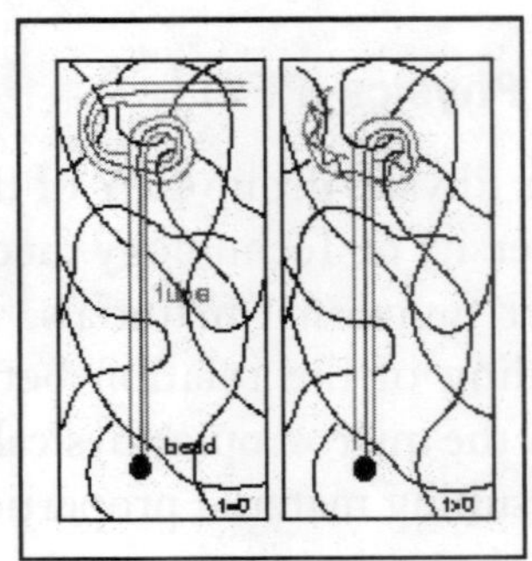

2.3 Theoretical and Polymer Physics

In our group, theoretical research is carried out on polymers and other soft materials. The central issue in this research is the understanding of the relation between the molecular and microstructure of materials, the microscopic physical processes bringing about these structures, and the resulting material properties. In short: the structure-dynamics-property relation. Methods from various areas of theoretical physics are used: statistical and soft-matter physics, solid-state quantum theory, but also large-scale computer simulations from the electronic to the mesolevel length- and timescales.

Research Area and Mission

- *Research area*: The understanding and prediction, on the basis of the underlying fundamental physics, of the triangle relation between (i) the molecular and mesoscopic structure, (ii) the dynamics of structure development, and (iii) the material properties, of polymers and related materials.
- *Mission*: To perform theoretical research at an internationally outstanding level in the field of polymers and other disordered materials, and thereby to have a forefront role in a multidisciplinary materials-science and materials-technology environment.

Collaborations and Embeddings

The group closely collaborates with other groups of the department, and with many polymer chemists, material scientists, biologists and computer scientists at the TU Eindhoven or elsewhere. An important part of the research is embedded in the Eindhoven Polymer Laboratories (EPL), the Eindhoven Center for NanoMaterials (cNM), the Dutch Polymer Institute (DPI), the Dutch Research School of Theoretical Physics (DRSTP), the local Netwerk Theoretische Fysica (NTF), and the Eindhoven Institute for Complex Molecular Systems (ICMS).

2.4 Theoretical Polymer Physics (PFY)

In the Theoretical Polymer Physics group PFY of the department of Applied Physics at Eindhoven University of Technology theoretical research is carried out on polymers and other functional materials. The central issue in this research is the understanding of the relation between the molecular and microstructure of materials, the microscopic physical processes bringing about these structures, and the resulting material properties. In short: the *structure-*

property relation. Methods from various areas of theoretical physics are used: statistical physics, percolation theory, solid state quantum theory, but also large-scale computer simulations. The group closely collaborates with other groups of the department, and with polymer chemists, material scientists and computer scientists from the departments of Chemical Engineering, Mechanical Engineering and Mathematics and Computing Science. An important part of the research is embedded in the Dutch Polymer Institute (DPI).

2.5 Theoretical Polymer Density Map

Adjusting the length and proportion of two kinds of polymer projecting from a minuscule bead would lead to a variety of geometric shapes, such as the one shown in this flattened density map of a spherical surface, according to computer simulations.

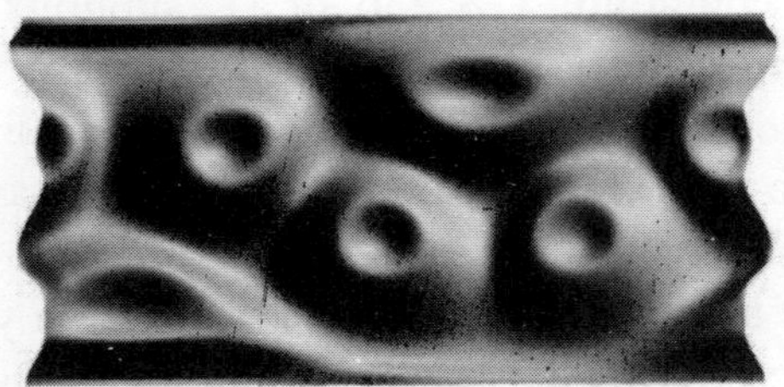

Fig. 2.1: Polymers to Polyhedra.

2.6 Polymers in a Vacuum May Yield Valuable Data

A theoretical analysis of polymer behavior suggests that large molecules should behave very differently when they are in a vacuum than when in solution. The finding suggests that with better understanding of the physics involved, analysis techniques that operate in vacuums—such as mass spectrometry and studies of organic molecules in space—could yield far more information than they do now. "Mass spectrometry of macromolecules was a huge breakthrough. But you're only getting one number from it, the mass of the molecule, and that's that," said Joshua Deutsch, professor of physics at the University of California, Santa Cruz.

In a paper published in the November 16 issue of *Physical Review Letters*, Deutsch analyzed the behavior of large molecules in a vacuum. He said that when long, chainlike molecules are released into a vacuum, they should oscillate in characteristic ways according to their makeup. By measuring those oscillations with electromagnetic sensors, scientists might someday be able to reconstruct details of the chemical structures of the molecules.

Many important biological molecules are macromolecules, including proteins, carbohydrates, fats, and nucleic acids like DNA. In the natural world, these polymers—long chains of repeating units connected by strong chemical bonds—are always in solution in our cells. So medical researchers tend to study them in solution. As anyone who has tried to run in a pool can attest, water restricts movement. Tiny structures such as biological macromolecules lie suspended in a cell's fluid, like spaghetti in molasses, Deutsch said. Weak chemical attractions between the solution and the molecule often enlarge the molecule, holding it in extended positions.

But in a vacuum, Deutsch realized, there is no medium to dampen movement—only the internal forces within the molecule itself. He found that this internal friction leads to very different behavior from that of a molecule in solution. Deutsch wrote a computer simulation of the physics involved in simplified molecules containing 128, 256, or 512 subunits arranged in a single chain. When he ran the model, he was amazed to see that points on the chain would periodically get closer, then farther apart, and continue in prolonged oscillations similar to the way the sides of a bell vibrate as it rings.

"It took me completely by surprise," Deutsch said. "No one had ever looked into this before, and based on what you see in solution you wouldn't expect the system to just ring and ring like that." After more analysis, Deutsch realized that the answer could be explained by a simple concept several centuries old. "Turns out it's just Galilean invariance," Deutsch said, referring to the principle that objects moving in a vacuum don't slow down. This means that each subunit in a macromolecule feels pushes and pulls from its neighbors, but the overall movement of the molecule is not restricted by a solvent.

"This gives rise to a far less efficient damping than you'd get in solution," Deutsch said. The oscillations are in the radio-frequency range and are strong enough, perhaps, to be detected in the vacuum chamber of a mass spectrometer, Deutsch said. First developed in the early 20th century, mass spectrometers were for years confined to studying atoms and small molecules. Mass spectrometers work by accelerating molecules over a specified distance and measuring how long the journey takes. With good enough instruments, this reveals the mass of the particle.

Mass spectrometry of macromolecules at first faced a major obstacle: no one could develop a way to vaporize them intact. Then, in the early 1980s, a young Japanese electrical engineer named Koichi Tanaka discovered a method by accident after making a faulty lab preparation. He followed up on the mistake with more research, eventually developing a workable method for which he won the 2002 Nobel Prize in chemistry.

In the new paper, Deutsch suggested that a mass spectrometer outfitted with the proper instruments could measure the oscillations from a vaporized macromolecule. By comparing the measurements against known samples and theoretical predictions, researchers might learn much more about the molecule than just its mass. Characteristics such as the arrangement of bonds, placement of charges, or atomic composition could possibly be determined. "For example, DNA, because it's a double helix, is much stiffer than a single chain," Deutsch said. "You ought to be able to measure that." For now, Deutsch is keeping things simple. His models look at single strands composed of identical subunits, with electrical charges acting only at each end of the chain. This is a far cry from typical macromolecules, which are a tangle of chemical bonds, contain many subunits in varying sequences, and have charges dispersed along their length. But Deutsch believes in starting simply. "I could add that all in right now, but what would I learn?" he said. "I want to work out the basics of this system. Then the experimentalists can take it from there."

3

Applied Polymer Science and Technology

3.1 Applied Polymer Science and Plastics CPU

Thistlebond Polymers are manufactured using *Advanced Polymer Science* that can offer many many advantages over comparative products and the following unique benefits to the end users!!

Feature	*Benefit*
Increased Chemical Resistance	Can be many % points Higher Than Competitors Products
Increased Temperature Resistance	Can be many % points Higher Than Competitors Products
Increased Life Expectancy	As APPLICATION costs are ESCULATING the LONGER the material LASTS the LESS expensive it in REALITY is.
No Solvents	This is now being INSISTED upon by industry. Majority of products are 100% solids. WHAT YOU PAY FOR IS WHAT YOU GET!!
No Shelf Life	Don't INVEST in stock from the competition that can end up as being provided for against profits!
Multiple Application Capability	Multi Role Use of polymers on VARIOUS sub-straits. MAJOR cost savings
Increased Flexibility	Less Likelihood of CRACKING and FAILURE associated with competitors polymer products
Chemical Adhesion	Not adhesion with SOLVENTS as most competitors
Extreme Abrasion Resistance	Can be up to 5 times better than the competition
Highest Hardness	Can be over 20% higher than the next comparable product
VOC Regulations	All Products Apply/SAFE TO USE
Free Technical Support	As Required

(*Contd.*)

(*Contd.*)

Feature	*Benefit*
Most Extensive R&D in the world	Ensure that you GE the RIGHT product at the RIGHT price for all your applications
Multiple Accreditations	Many Many Test Accreditations—BUY WITH CONFIDENCE
Multiple Referrals	Some Where Some One has carried out a successful application that you may now want to do!
Quick Delivery	Distributors/Agents in most countries
Competitively Priced	Buy with Confidence
Increased Production Capability	In-Situ Repairs Mean LESS Downtime
Thin Coats—Usually Only Two	More Flexible Product—More Able To Withstand Thermal Cycling Effects—Retains Impact Resistance—Retains High wear Capabilities—LESS EXPENSIVE TO APPLY—Products can go further than most competitor products. Batch Process Checked at all times and retained for future investigation if required
Quality Assurance ISO 9001: ISO 4001:	Buy With Assurance
Permanent repairs	Longer lasting investment in product and applications costs—less capital outlay—order with confidence!

Introduction

The Polymer Science and Plastics CPU introduces students to the science and processes required for plastics manufacturing. These processes include injection molding, casting, vacuum forming, rotational molding, blow molding, and extrusion. The included activities and multimedia instruction explore the many industries that are impacted by plastic and the careers that they host. The Polymer Science and Plastics CPU is available in 10 and 15-hour formats.

Featured Careers

- Biological scientist
- Cytotechnologist
- Fashion designer
- Agricultural scientist
- Machinist and tool programmer
- Auto body repairer.

Skill Development

Technology Topics and Skills

- Career and occupational outlook

- Academic requirements
- Shop safety practices
- History of polymers/plastics
- Plastics manufacturing processes
- Natural polymers
- Latex, rubber, vulcanization
- Synthetic polymers
- Medical, agricultural, and textile applications
- Vacuum forming, injection molding, rotational molding, plastic welding, and fabrication

Academic Skills

- Reading
- Writing
- Science—chemistry
- Physical science
- Communications
- Research
- Teamwork
- Problem solving
- Resource allocation
- Technological literacy
- Skills event alignment
- TSA event alignment
- FFA event alignment.

What's Included

- Applied Technologies Polymer Science & Plastics courseware CD
- Applied Technologies Polymer Science & Plastics video CDcs
- Applied Technologies Polymer Science & Plastics Student Guide
- Applied Technologies vacuum former
- Injection molder and accessories
- Screwdriver molds
- Wheel molds
- Styrene pellets
- Biodegradable plastics textbook
- DNA construction kit
- Rotary grinder
- Car molds
- Sheets of plastic

- Axles
- Stopwatch
- Plastics technology video
- Plastics production techniques video
- Hydroponics video
- Miscellaneous materials and consumables
- Instructor's Overview
- Media Cruiser

3.2 Center for Applied Polymer Research

Department of Macromolecular Science and Engineering: Case Western Reserve University

The Center for Applied Polymer Research (CAPRI) directed by Professor Anne Hiltner has gained national recognition for the cutting edge research in a wide spectrum of topics. Located in the Department of Macromolecular Science and Engineering of Case Western Reserve University, CAPRI currently consists of about twenty graduate students, two postdoctoral research associates, and several undergraduates.

Combining state-of-the-art facilities and student education with industrial collaboration in applied polymer science and engineering has been the formula for success in research topics spanning from hierarchical structures to nanolayered materials.

Mission

CAPRI's mission is to carry out interdisciplinary applied and basic research on structure-property relationships in polymeric materials of interest to various industries. CAPRI will serve as an intellectual center for the development of new materials concepts and new analytical techniques through the traditions of quality advanced graduate education. CAPRI will foster university-industry interactions in order to provide identifiable returns including educated people and knowledge to its sponsors as it develops into an internationally recognized center. The Center for Applied Polymer Research carries out research on four different levels: polymer blends and alloys, structural composites, processing of layered materials and structures, and polymers for biomedical applications. Through these initiatives, the Center serves the growing need for complex materials systems by developing new materials concepts and innovative analytical techniques for its industrial partners. The Center has also identified educational requirements for scientists and engineers of the 21st century and incorporates these requirements into graduate-level curricula.

3.3 Research Areas

Our interests focus on the relationships between hierarchical structure and mechanical function of polymeric materials. Hierarchical structure in biocomposite systems such as in collagenous connective tissue has many scales or levels, and there are highly specific interactions between these levels. We were among the first to demonstrate that the hierarchical architecture is designed to accommodate a complex spectrum of mechanical property requirements. The hierarchical structure-property relationships have been described in several soft connective tissues: tendon, intestine and invertebral disc. Numerous levels of organization are found with highly specific interconnectivity and unique architectures designed to give the required spectrum of properties for each oriented composite system. From these lessons in biology, the laws of complex composite systems for functional macromolecular assemblies are considered.

The other area of interest focuses on exploring the structure-processing-property relationships in polymeric materials. Pioneering contributions have been made in understanding the connections between hierarchical structure and irreversible deformation and damage processes, and fracture of polymer blends and composites. New insights have been developed into mechanisms of compatibilisation and toughening of polymer blends. We have pioneered in the development of effective industry/academic research cooperation. As an example, with Dr. S. P. Chum of The Dow Chemical Company, a unique and comprehensive study of the structure-property relationships of metallocene-catalysed ethylene-octene copolymers was carried out, which has led to a definitive theoretical understanding of the connections between microstructure, morphology and mechanical properties. This work has, in turn, produced predictive models which have assisted the development of a new commercially-successful family of ethylene-based elastomers. Motivated by the need for new processing technologies for creating engineered microstructures of incompatible polymers, ground-breaking studies have been undertaken to explore the unique advantages that can be achieved with microlayering coextrusion. This layer-multiplying technology permits continuous processing of sheet or film with hundreds or thousands of alternating layers of two or more polymers. With this technology, we have created engineered microstructures with unique electrical, mechanical and barrier properties. With E. Baer, the hierarchical structure-function relationships in collagenous tissues have been explored. Numerous levels of organization are found with highly specific interconnectivity designed to produce the particular spectrum of properties for each oriented composite system. A further area of interest with J. M. Anderson is in understanding mechanisms of biocompatibility and biodegradation of biomaterials with a view to enhancing their biostability.

3.4 General Information

The internationally oriented Master Course in Applied Polymer Science is specially designed to teach in English language in order to attract international as well as German students. A requirement for entering the course is a Bachelor Degree in physics, pharmacy or chemistry. Students having a Bachelor Degree in engineering are asked to have a look to the new curriculum that you can find on this homepage. As you can see there are high requirements in chemistry, physics and mathematics. So, if you have a Bachelor Degree in engineering and you wish to enter the course please contact responsible persons via email.

After successful completion of the Master examination, the Faculty of Natural Sciences II—Chemistry and Physics at Martin Luther University awards the academic degree *Master of Science (M. Sc.):*

Duration	*Mode*	*Level*	*Application to*	*Points 2007*	
				Median	*Final*
1 year 2 years	Full-Time Part-Time	9	Dept. of Mechanical & Plastics/Polymer Engineering, AIT	N/a	N/a

The overall philosophy of the course is to address the need to provide additional skills to professional engineers and technologists currently working in polymer related, or hoping to work in polymer related industries in Ireland. The course will take advantage of the participants' academic and professional background. Hence, a fast learning curve will be assumed and efforts will be made to provide breadth and depth to the various areas of applied polymer technology under study. Special emphasis will be placed on leading edge technologies and industrial issues.

Minimum Entry Requirements

Honours degree in a relevant discipline such as polymer technology or polymer engineering, or an appropriate equivalent level-8 qualification. Additional industrial experience may be required depending on the degree qualification.

Course Subjects

- Engineering and Product Design, Polymer Materials, Computer Aided Design and Analysis, Polymer Processing Technology, Project.

- ❖ Plus four subjects from(electives offered subject to viable numbers)
- ❖ Project Management, Polymer Materials, Biomedical Science and Technology, Regulatory Affairs and Validation, Biomedical Materials, Packaging and Sterilisation.

1st Semester

The first semester is designed to give a broad introduction to polymer science. Also the lab capabilities of the students will be improved. Furthermore, the knowledge in physics and chemistry, respectively, towards an improved level is intended.

2nd and 3rd Semester

The second semester will improve the knowledge and experimental skills of the students in the field of polymer physics and polymer chemistry.

In the third semester, the students will be introduced to polymer research, including a 150 hours project work. Then, they can select between four specializations: polymer physics, polymer chemistry, polymer engineering, and bio-related polymers.

4th Semester

This term is to carry out your Master's Thesis. It should be written in English language, an oral defense has to be passed.

Usually, the research will be done at the Departments of Physics or Chemistry. Polymer related topics can also be supervised in other Departments at Martin-Luther-University, or outside of it (e.g. Max-Planck-Institute, Fraunhofer Insitute).

3.5 Using Advanced Polymer Applications

Polymer Coating and Repair Materials

"Unique Polymer Systems is dedicated to bring both the end user and distribution companies that it works closely with, new and innovative products that fill a hole in the market place with regard to the 'normal' products." To compliment this "supply" service we also offer in the UK an APPLICATION service that consists of the following advantages to the end user and Distributor Company.

To achieve MAXIMUM life from any polymer coating or repair material certain steps MUST take place.

Don't Accept Less from Your Applicator

In the first place you need to choose a company that has an extensive background in polymer engineering as well as a good knowledge of both civil and mechanical engineering disciplines.

It is well know that over 70 per cent of the cost of most polymer type applications is in the cost of SURFACE PREPARATION and APPLICATION of selected product. "Over 95 per cent of all application failures in the field can be traced back to poor surface preparation" Unique Polymers is a polymer provider in its own right. The selection of the RIGHT polymer is always offered as selected from exacting specifications from a highly qualified manufacturer's base of world class companies. Unlike other applicator companies who have agreements with a specific manufacture (even though they may NOT have the best product for the application)—we can offer an impartial selection criteria from many manufacturing companies world wide.

Access to the best chemists who can assist in making the right choice of product for each application "right first time" If there is a requirement for testing in both laboratory and real life situations we can offer to do this for you to.

3.6 Applied Polymer Science—International M.Sc. Degree Course

- *Degree*: Master of Science
- *Duration*: 4 semesters

Admission Requirements

- *Bachelor Degree* (B.Sc.) in *Chemistry* or *Physics*. Other degrees in materials science, biochemistry, biophysics etc. can be recognized on a case by case basis.
- *Language Requirements*: German students need at least the UNIcert II. Foreign students need to have the TOEFL test.

Admission Procedure and Deadlines

Admission procedure and deadlines: There is a board for the selection of applicants. The matriculation is for the winter semester only. Deadline for application is June 15th of every year.

Studying at the Faculty II

With its 500-years-old history, the Halle University holds a prominent position in German and Euro-pean social and natural sciences. It has a long tradition in polymer science, e.g. Nobel laureate Karl Ziegler was a professor at our university in the field of Technical Chemistry.

Characterization of the Masters Course and Job Profile

The international Master Course of Applied Polymer Science was first offered in the 2001/2002 winter semester as a cooperation of the chemistry, physics, and engineering departments. It is designed in-terdisciplinarily and is focused on polymer chemistry and polymer physics with some aspects of poly-mer engineering. Based on sound natural-scientific fundamentals, theoretical and experimental skills in the field of polymers are developed. The aim is to enable chemists and physicists to successfully apply their knowledge and skills in academia as well as in industry (e.g. in the Halle/Leipzig, Bitterfeld area, BASF, Bayer, DOW). As an international course, it also aims at job opportunities in High-Tech regions worldwide, in developing countries as well as East European countries. Existing contacts with national and international companies and universities are used and consolidated for the training of students. The university campus is located in Halle and provides excellent conditions for living and studying.

The Course

The duration for the course is four semesters, including the six month period for the Master thesis. The curriculum, lectures, lab courses and seminars as well as the examination procedures ensure that the course can be completed with in these four semesters. Upon matriculation the students have to select between the specializations in polymer physics and polymer chemistry, respectively. The master course has 120 credit points. After getting the basics of polymer chemistry, polymer physics, and polymer engineering in the first and second semester, the third semester provides skills in four differ-ent fields: i) Advanced Polymer Physics, ii) Advanced Polymer Chemistry, iii) Advanced Polymer Engi-neering, and iv) Bio-related Polymers. The Master thesis can be carried out at the university or in in-dustry. All teaching is done in English language. Classes in German language are offered additionally. There is no tuition fee.

Specialists in the design of polymer systems and products for engineering applications, including:

- vibration control
- marine engineering
- sub-sea protection
- acoustic signature management.

With a range of innovative and unique products, such as:

- rho-c materials (acoustic windows)
- castable syntactic foams for controlled buoyancy
- anechoic and transmission loss coatings
- shock mitigation systems
- broad band damping materials

APT provides effective solutions to material problems. Our web site gives a brief introduction to our capabilities.

3.7 Polymer Design and Engineering

At the heart of APT's design and engineering capability is our detailed knowledge of the dynamic properties of polymers and how they can be manipulated to give innovative solutions to acoustic and vibration control applications. By combining polymer chemistry and component geometry, systems can be tailored to meet acoustic and other performance specifications, such as operational frequency range, echo reduction, insertion loss and environmental conditions. The first figure shows the dynamic modulus and damping of a typical viscoelastic polymer. By changing the polymer chemistry, both dynamic modulus and damping can be changed to give effective damping at a wider frequency or temperature range to tune the damping to a specific frequency to increase the amount of stored energy to minimise heating during cyclic loading. The second figure shows a material optimised for broad band damping.

3.8 Acoustic Test Facility

APT has its own acoustic pulse tube test facility, used as a development tool, to allow quick measurement of prototype systems for comparison with customer specifications, and for internal quality control testing of acoustic materials. The water-filled tube is 4.7m long, and can be used to measure echo reduction, insertion loss and fractional power dissipation as a function of frequency (3-9 kHz) and temperature (1-30°C). The tube can also be pressurised to simulate the effect of hydrostatic pressure on acoustic performance (atmospheric pressure to 600 psi—equivalent to a depth of 400 m). An impulse signal is projected into the tube from one end, with side hydrophones capturing the signal passing through, and reflected from, the test piece. Data is converted into the frequency domain, using Fast Fourier Transforms, allowing measurement of acoustic performance. The pulse tube can also be used for multiple pressure cycling to test bond integrity, to examine material deformation (creep and set) under sustained loading and to investigate the effect of repeated excursions in pressure on acoustic performance.

3.9 Manufacturing

APT is a specialist in the design and manufacture of a wide range of synthetic and natural rubbers, polyurethanes, epoxides and thermoplastics. Materials and components are not originated in isolation, but are designed to be integrated into a full, working, manufacturable system. This provides for the complete involvement of mould tool design, component geometry, performance specification, processing requirements, so that the product is not only functionally suitable, but is also viable for manufacture.

With a well-equipped manufacturing plant, APT can handle all aspects of the process from design through to production. All manufacturing procedures are underpinned by our ISO 9001:2000 accredited quality system.

3.10 Acoustic and Underwater Materials

APT supply "off-the-shelf" and custom acoustic materials and materials for use underwater. Materials are designed to meet specific acoustic, operational and environmental requirements. Our capability includes acoustic tiles and fairing strips, rho-c materials and encapsulation of electronics. Please contact us for further details.

- ❖ Rho-c materials (acoustic windows).
- ❖ Castable syntactic foams.
- ❖ Anechoics.
- ❖ Transmission (Insertion) Loss systems.
- ❖ Decouplers.
- ❖ Thixotropic transmission loss sealants.

3.11 Advanced Material Technology

Conventional acoustic materials are produced using single relaxation polymers. The dynamic properties of single relaxation polymers are generally sensitive to any variation in operational temperature and frequency. APT have developed and supply a range of broad band damping materials, which have specific mechanical, dynamic and acoustic properties which are considerably less sensitive to variations in temperature.

The materials are synergistic co-polymer blends, and Dynamic Mechanical Thermal Analysis shows effectively a single relaxation material with a broad band transition zone and damping coefficient.

The performance of these materials are desensitised to operational conditions, such as temperature and frequency. The materials can readily be moulded into intricate geometries for applications where material properties and component geometry combine to determine system performance.

This gives significant advantages over conventional acoustic materials, where performance is optimised for a narrow temperature range, and deteriorates outside this range. As the materials have been designed to give excellent reproducibility and process capability they can offer significant economies over conventional acoustic materials, including the possibility of using a single broad band material component to provide both echo reduction and transmission loss.

Physical properties of the broad band materials are considerably higher than those of conventional acoustic materials, giving enhanced survivability in service. Typical tensile strength of a broad band material is 18.5MPa, compared with 3.5 MPa for a polyurethane-based anechoic.

A wide range of properties can be achieved with these broad band damping materials, and they can be adapted to meet specific performance requirements, such as low creep, resilience, and dynamic fatigue resistance.

3.12 Expert Advice

As the leader in the development of polymer technology, we can provide the precision testing, material analysis, polymer characterization, compound recommendations, ideas for process improvement, and advice on how to bring your product to market more efficiently. No matter what your product is, it just makes sense to trust your material and processing needs with the people who helped invent the technology: PDI.

Our Polymer Technology expertise can help you with:

- Product lifecycle prediction,
- Isolation and resolution of quality control issues,
- Correction of manufacturing and process issues,
- Accelerated weathering and outdoor exposure studies,
- Accredited physical properties testing,
- Material analysis and identification,
- Process design and process engineering support in chemical analysis and microscopy,
- Characterization of biomaterials,
- Polymer delivery systems for pharmaceuticals,
- Advanced rheology issues, and
- Failure analysis and prevention.

4 Polymer Rheology

4.1 Rheology

Rheology is the study of the flow of matter: mainly liquids but also soft solids or solids under conditions in which they flow rather than deform elastically. It applies to substances which have a complex structure, including muds, sludges, suspensions, polymers, many foods, bodily fluids, and other biological materials. The flow of these substances cannot be characterized by a single value of viscosity (at a fixed temperature—instead the viscosity changes due to other factors. For example ketchup can have its viscosity reduced by shaking, but water cannot. Since Isaac Newton originated the concept of viscosity, the study of variable viscosity liquids is also often called *Non-Newtonian fluid mechanics*. The term *rheology* was coined by Eugene C. Bingham, a professor at Lafayette College, in 1920, from a suggestion by a colleague, Markus Reiner. The term was inspired by the quotation mistakenly attributed to Heraclitus, (actually coming from the writings of Simplicius) *panta rei*, "everything flows". The experimental characterisation of a material's rheological behaviour is known as *rheometry*, although the term *rheology* is frequently used synonymously with rheometry, particularly by experimentalists. Theoretical aspects of rheology are the relation of the flow/deformation behaviour of material and its internal structure (e.g. the orientation and elongation of polymer molecules), and the flow/deformation behaviour of materials that cannot be described by classical fluid mechanics or elasticity.

Scope

In practice, rheology is principally concerned with extending the "classical" disciplines of elasticity and (Newtonian) fluid mechanics to materials whose mechanical behavior cannot be described with the classical theories. It is also concerned with establishing predictions for mechanical behavior (on the continuum mechanical scale) based on the micro- or nanostructure of the material, e.g. the molecular size and architecture of polymers in solution or the particle size distribution in a solid suspension. Materials flow when subjected to a stress, that is a force per area. There are different sorts of stress and materials can respond in various ways, so much of theoretical rheology is concerned with forces and stresses.

<table>
<tr><td rowspan="4">Continuum mechanics</td><td rowspan="2">Solid mechanics or strength of materials</td><td colspan="2">Elasticity</td></tr>
<tr><td>Plasticity</td><td rowspan="2">Rheology</td></tr>
<tr><td rowspan="2">Fluid mechanics</td><td>Non-Newtonian fluids</td></tr>
<tr><td colspan="2">Newtonian fluids</td></tr>
</table>

Rheology unites the seemingly unrelated fields of plasticity and non-Newtonian fluids by recognizing that both these types of materials are unable to support a shear stress in static equilibrium. In this sense, a plastic solid is a fluid. Granular rheology refers to the continuum mechanical description of granular materials.

One of the tasks of rheology is to empirically establish the relationships between deformations and stresses, respectively their derivatives by adequate measurements. These experimental techniques are known as rheometry and are concerned with the determination with well-defined *rheological material functions*. Such relationships are then amenable to mathematical treatment by the established methods of continuum mechanics.

The characterisation of flow or deformation originating from a simple shear stress field is called shear rheometry (or shear rheology). The study of extensional flows is called extensional rheology. Shear flows are much easier to study and thus much more experimental data are available for shear flows than for extensional flows.

Rheologist

A rheologist is an interdisciplinary scientist who studies the flow of complex liquids or the deformation of soft solids. It is not taken as a primary degree subject, and there is no general qualification. He or she will usually have a primary qualification in one of several fields: mathematics, the physical sciences, engineering, medicine, or certain technologies, notably materials or food. A small amount of rheology may be given during the first degree, but the professional will extend this knowledge during postgraduate research or by attending short courses and by joining one of the professional associations.

Applications

Rheology has applications in engineering, geophysics, physiology and pharmaceutics. In engineering, it affects the production and use of polymeric materials, but plasticity theory has been similarly important for the design of metal forming processes. Many industrially important substances such as concrete, paint and chocolate have complex flow characteristics. Geophysics includes the flow of lava, but in addition measures the flow of solid Earth materials over long time scales: those that display viscous behaviour, e.g. granite, are known as rheids. In physiology, many bodily fluids are have complex compositions and thus flow characteristics. In particular there is a specialist study of blood flow called hemorheology. The term biorheology is used for the wider field of study of the flow properties of biological fluids.

Elasticity, viscosity, solid and liquid-like behavior, plasticity: One generally associates liquids with viscous behaviour (a *thick* oil is a viscous liquid) and solids with elastic behaviour (an elastic string is an elastic solid). A more general point of view is to consider the material behaviour at short times (relative to the duration of the experiment/application of interest) and at long times.

Liquid and solid character are relevant at long times: We consider the application of a constant stress (a so-called *creep experiment*):

- if the material, after some deformation, eventually resists further deformation, it is considered a solid
- if, by contrast, the material flows indefinitely, it is considered a liquid.

By contrast, *elastic and viscous* (or intermediate, viscoelastic) behaviour is relevant at short times (*transient behaviour*)

We again consider the application of a constant stress:

- if the material deformation follows the applied stress, then the material is purely elastic,
- if the deformation increases linearly at constant stress, then the material is viscous,
- if neither the deformation with time, nor its derivative (*deformation rate*) follows the stress, the material is viscoelastic.

Plasticity is equivalent to the existence of a yield stress: A material that behaves as a solid under low applied stresses may start to flow above a certain level of stress, called the *yield stress* of the material. The term *plastic solid* is often used when this plasticity threshold is rather high, while *yield stress fluid* is used when the threshold stress is rather low. There is no fundamental difference, however, between both concepts.

Dimensionless Numbers in Rheology

Deborah Number

When the rheological behavior of a material includes a transition from elastic to viscous as the time scale increase (or, more generally, a transition from a more resistant to a less resistant behavior), one may define the relevant time scale as a relaxation time of the material. Correspondingly, the ratio of the relaxation time of a material to the timescale of a deformation is called Deborah number. Small Deborah numbers correspond to situations where the material has time to relax (and behaves in a viscous manner), while high Deborah numbers correspond to situations where the material behaves rather elastically.

Note that the Deborah number is relevant for materials that flow on long time scales (like a Maxwell fluid) but *not* for the reverse kind of materials (like the Voigt or Kelvin model) that are viscous on short time scales but solid on the long term.

Reynolds Number

In fluid mechanics, the Reynolds number is a measure of the ratio of inertial forces ($v_s \rho$) to viscous forces (μ/L) and consequently it quantifies the relative importance of these two types of effect for given flow conditions. Under low Reynolds numbers viscous effects dominate and the flow is laminar, whereas at high Reynolds numbers inertia predominates and the flow may be turbulent. However, since rheology is concerned with fluids which do not have a fixed viscosity, but one which can vary with flow and time, calculation of the Reynolds number can be complicated.

It is one of the most important dimensionless numbers in fluid dynamics and is used, usually along with other dimensionless numbers, to provide a criterion for determining dynamic similitude. When two geometrically similar flow patterns, in perhaps different fluids with possibly different flow rates, have the same values for the relevant dimensionless numbers, they are said to be dynamically similar.

Typically it is given as follows:

$$Re = \frac{\rho v_s^2 / L}{\mu v_s / L^2} = \frac{\rho v_s L}{\mu} = \frac{v_s L}{v}$$

where

v_s mean fluid velocity, [m s^{-1}]
L characteristic length, [m]
μ (absolute) dynamic fluid viscosity, [N s m^{-2}] or [Pa s]
v kinematic fluid viscosity: $v = \mu/\rho$, [m^2 s^{-1}]
ρ fluid density, [kg m^{-3}].

Applied Rheology

Applied Rheology covers the science of the deformation and flow of soft matter, with special interest in experimental and computational advances in the characterization and understanding of complex fluids, including their nonequilibrium dynamic and structural behaviors. The journal offers insight into both phenomenological and molecular theories, instrumentation, the study of diverse materials (such as polymers, rubber, paint, glass, foods, biological materials) and a wide range of practical applications.

We are inviting the contribution of manuscripts that present new techniques or give an overview of existing experimental, computational or theoretical methods in the field of applied rheology. The contributions should be at a review level with a strong tie to present or future applications as well as understandable to the non-researcher. Applied Rheology also serves as the newsletter for several rheological societies worldwide.

4.2 Meaning of Rheology

Rheology, the study of the flow and deformation of matter, is an old discipline undergoing a renaissance. In its widest sense, it includes classical fluid mechanics and elasticity which treat the flow of Newtonian liquids, such as water, and small deformations of hard solids, such as wood and steel.

The use of the special term "rheology" for these subjects alone would not be justified, since they have been extensively studied for more than 170 years, and are an accepted part of the curriculum in most universities.

In practice, the word "rheology" normally refers to the flow and deformation of "non-classical" materials such as rubber, molten plastics, polymer solutions, slurries and pastes, electrorheological fluids, blood, muscle, composites, soils, and paints. These materials can exhibit varied and striking rheological properties that classical fluid mechanics and elasticity cannot describe. Though the word "rheology" was coined in 1929, the rapid development of the subject began 20 years later.

Today, there are societies of rheology in Argentina, Australia (est. 1983), Austria (est. 1969), Belgium (est. 1974), Britain (est. 1940), Canada (est. 1982), China, Czechoslovakia (est. 1968), France (est. 1955), Germany (est. 1951), Greece (est. 1996), India, Israel, Italy, Japan (est. 1973), Korea, Latvia, Mexico, The Netherlands (est. 1951), the Nordic countries (est. 1956), Poland (est. 1997), Portugal (est. 1997), Romania, Slovenia, the former Soviet Union (est 1964), Spain (est. 1982), Switzerland (est. 1993), and the United States (est. 1929). There is also a European Society of Rheology, an International Society of Biorheology, a Japanese Society of Biorheology and an International Society for Clinical Hemorheology and Microcirculation. Nine journals are devoted to rheology: *Rhéologie* (formerly *Les Cahiers de Rhéologie), Clinical Hemorheology and Microcirculation*, *Biorheology*, *Journal of Non-Newtonian Fluid Mechanics*, *Journal of Rheology*, *Korea-Australia Rheology Journal*, *Nihon Reoroji Gakkaishi*, *Rheologica Acta*, *Applied Rheology* and *Rheology Abstracts*. The latter lists more than 100 other journals which publish articles on a variety of rheological interests. *Polymer Engineering & Science* devotes six issues per year to Polymer Processing & Rheology. Every four years, the latest developments are archived in the *Proceedings of the International Congress on Rheology*.

Rheological research involves activities in and draws on knowledge from biophysics, chemical engineering, chemistry, computer science, electronics, engineering mechanics, materials science, mathematics, mechanical engineering, medicine and physics among others. Rheology's interdisciplinary nature stems from the variety of materials investigated and the many new questions that must be answered.

In classical mechanics, material properties are determined when one or two constants are given (viscosity or elastic modulus); the basic "constitutive equations" governing the behavior of each small material element are known. Further research often involves applying these equations, with the momentum conservation equation, to increasingly complicated flows or deformations.

In rheology, on the other hand, the constitutive equations for most materials are unknown, and frequently involve unknown functions; the form of the equations for visco-elastic materials is so different from the classical forms that one must combine continuum mechanics with molecular theory and test the predictions with measurements from a variety of flows or deformations. The molecular theory requires statistical mechanical modeling with computer analysis. The measurements require inventing new instruments for determining nonlinear viscoelastic properties, stress distributions, elastic recoil and flow birefringence. While intuition suggests that structural characteristics like branching should hinder the flow of molecules, it is hard to incorporate such effects with purely continuum arguments. Recent molecular simulations at the Rheology Research Center have helped explain the effect of molecular structure on rheology from first principles.

The challenge of relating molecular structure or morphology to the macroscopic rheological properties of materials is fundamentally important, whether for molten plastics, polymer solutions, suspensions or composites. Meeting this challenge deepens our understanding of rheological properties, guides improvements in processing technology and helps us produce new materials for specific applications.

Polymer rheology is perhaps the most extensively and successfully studied field in rheology. Considerable progress has been made in developing constitutive equations from molecular models and also in developing special rheometers for testing theoretical predictions in the laboratory. Future rheological research will combine constitutive equations with heat flow equations to solve more complicated flow problems to assist the design of polymer processing equipment.

Rheology thus involves, to an unusual degree, a combination of academic and scholarly activities at an advanced level with the possibility of practical applications in a wide range of industrial situations. These include commercial processing of plastics and rubber, textiles, paper, oil and paints, foodstuffs, adhesives and composites.

4.3 Rheology Solutions

Materials Testing for the Polymer Processing Industry

Polymer processing includes unit operations such as compression, injection and blow moulding and extrusion through a variety of dies. Specifically for the polymer industries, materials characterisation is of critical importance (including fluid related issues like sedimentation, time related structural decay or build-up and post-cure issues like strength of compression or extension) for prediction and management of sharkskin, melt fracture, calendering, die swell, melt homogeneity and short- and long-term dimensional stability of extrudates or moulded parts. These processes depend on fluid rheological parameters such as viscosity, visco-elasticity, creep and recovery, and solid material properties such as behaviour under various compressive and extensional loading conditions. Extensional flow properties dominate processes where stranding occurs and influences time, quality and energy requirements.

The Rheology Solutions applications laboratory has state of the art equipment, capable of measuring all of the above parameters, and of providing interpretation of the results where necessary. Laboratory equipment includes highly sensitive, specialised, modern instrumentation and sensor systems, for measuring complex or difficult fluids such as those with very low viscosity, or those with a highly settling solid phase, or liquids with significant extensional or visco-elastic properties. Solid material properties according to various international standards can also be monitored.

Information such as shear viscosity curves, thixotropic behaviour and flow curves may be obtained for interested clients. Additionally, of importance to the industry, when scientifically assessing changes in flow behaviour due to temporal or ingredient issues, viscoelastic moduli as a function of temperature (up to 250°C) shear rate or strain can be assessed. Structural changes under very low shear (like in the case of storage, gravity settling, compression moulding etc) or very high shear (pumping, extrusion, mixing, injection moulding etc) are possible using the advanced equipment and sensors available at the laboratory of Rheology Solutions. These measurements may be obtained as a function of temperature, solids density, or to monitor the effects of changes in additive/ingredient species or concentration. With this information the scientist or engineer may change the process, or design new unit operations or products to maximise the potential benefits to be obtained from the physical properties of the processed fluid. Rheology Solutions is pleased to provide this information to interested parties, and to make our rheological expertise available to the polymer processing community. Our range of testing and interpretation services is further supported by customised workshops, training and seminars to cover the needs of individuals, businesses or industry sectors, covering introductory, intermediate or advanced theory and practice of rheology. This can be supplemented by application specific workshops.

A complete Contract Testing Information Kit specific to the Polymer Processing Industry is available by sending an email to info@rheologysolutions.com and typing "Contract Testing Kit for the Polymer Processing Industry" in subject area. You should receive your kit via post within 5-10 working days. The Contract Testing Request Form needs to completed and returned prior to any samples being sent. This will allow us to advise you within 48 hours if we require further information. We will then provide an overview of the work we will carry out including an estimate of the projected time scale and associated fee

4.4 Polymer Rheology: Course Information

Course Description (from the catalog): A systematic development of the principles and applications of the science of rheology. Reviews vector and tensor mathematics and Newtonian fluid dynamics. Develops the physical and mathematical nature of stress and deformations in materials. Covers the use of theory and application of rheological equations of state.

Credits: 3.0 *Lec-Rec-Lab:* (3-0-0).

Semesters Offered: On Demand (usually every Spring, MWF 3pm). Begining in Spring 2007 there will be an on-line section of CM4650.

Pre-requisites: (CM 3110 or MEEM 3210 or ENG 3200 or MY 3110 or CE 3600) and (MA 3520 or MA 3521 or MA 3530 or MA 3560) (i.e. fluid mechanics and differential equations).

Course Goals

1. Learn and practice vector and tensor mathematics that is necessary for understanding rheology.

2. Learn and understand constitutive theories and how they may be used to calculate flow fields and stress fields in flowing polymer systems.
3. Learn theoretical underpinings of rheometric techniques.
4. Learn and understand the wide variety of rheological response that is exhibited by non-Newtonian fluids.

4.5 Polymer Rheology and Flow Simulation

Rapra offer a wide range of rheological services to cater for all needs ranging from long/short term routine quality control to process design and flow analysis.

Specialities Include

Capillary Rheometry

- High Shear Viscometry
- Extensional Viscosity
- Pressure Coefficient of Viscosity
- NFT (No Flow Temperature)
- Melt Density

Controlled Stress/Rate Rotational Rheometry

- Low shear Viscometry
- Dynamic Oscillatory Rheometry
- Thermal Stability/degradation studies
- Change in viscosity as a function of temperature
- Yield Stress Analysis
- Creep/Creep Recovery

High Strain Rotational Rheometry (Rubber)

- Pre- & post-cure Dynamic Oscillatory Rheometry
- Cure Kinetics
- Variable Temperature Analysis

The flow simulation service provides accurate materials data for proprietary software packages such as Sigmasoft and Moldflow, with the facility to tailor testing to specific client needs.

- High & low Shear Viscometry
- Extensional Viscosity
- Pressure Coefficient of Viscosity
- Dynamic Oscillatory Rheometry

- PvT (Specific volume as a function of Pressure & Temperature)
- Thermal Diffusivity (Calculation of thermal conductivity)
- Specific Heat Capacity
- Ejection Temperature
- NFT (No Flow Temperature)
- Density
- Melt Density
- Cure kinetics (Rubber)

4.6 Polymer Analysis

Rapra's Polymer Analysis Team supports a wide range of industry sectors including healthcare, automotive, food packaging, recycling and oil. Our activities also include work for academia, the EC and UK Government Agencies (e.g. the Food Standards Agency). Rapra's UKAS accredited laboratories provide techniques to examine the chemical and physical properties of numerous materials and compounds including:

- Base polymers e.g. elastomers and thermoplastics
- Compounds e.g. rubbers, thermoplastics and thermoset formulations
- Additives e.g. fillers, plasticisers, antidegradants and curatives
- Polymer based products e.g. adhesives, paints, inks, lacquers, laminate films, plastisols and coatings
- Manufacturing sector products and components e.g. those used in the automotive, construction, electrical, medical and offshore industries

Rapra has a range of specialist services and techniques which can be commissioned independently or combined in a comprehensive consultancy project as required. Training courses are also available.

Analytical services typically support the following activities throughout the product lifecycle:

- Materials selection
- Product development
- Compound deformulation/formulation
- Service life prediction
- Quality assurance
- Process improvements
- Technical specification compliance
- Failure analysis
- Forensic investigation
- Expert witness service

4.7 Polymer Product Testing

How many times has a good plastics idea failed because the prospect of climbing the testing and certification mountain has seemed so difficult? A lack of technical back-up and the right kind of engineering skills can really deter the innovator and keep the innovation unproven. Rapra's world-class testing services, UKAS back-up and service make all of this process very much easier than you think. Our specialty is the client work we do on new products and testing regimes where no standards yet apply. We help design the rigs, jigs and fittings to your needs and we develop new testing methods most appropriate for your product. It's a one-stop package and we help in developing the whole job.

We have eighty years of history in getting it right. Rapra's involvement today in the main national and international standards-making bodies is second to none for plastics and rubber. Testing is our worldwide expertise and we are frequently called to give legal evidence in cases of product failure, liability or conflict of interest.

4.8 Cavity Transfer Mixing

The mixing of small quantities of low viscosity fluids into large volumes of much higher viscosity fluids is a problem encountered in most of the chemical industry; examples include (*i*) the addition of liquid colours in polymer and food processing; (*ii*) pharmaceutical, cosmetic and coating industries where a range of masterbatches are used to make speciality products.

In the polymer processing industry the distributive action needed to break up the globules of low viscosity masterbatches into finer droplets within the higher viscosity major phase is achieved by the use of distribtive mixing heads. Dispersive mixing however is difficult to induce since it requires elongational flows and depends on the flow conditions as well as the fluid properties (viscosity ratio of dispersed and continuous phase).

The area of research being addressed here is the use of a standard distributive mixing head (cavity transfer mixer (CTM)) and to improve its dipersive capabilities in a controlled and predictable manner. It involves the injection of a small quantity (1.0-5.0%) of a low viscosity colour at the inlet of the CTM into a high viscosity continuous polymer flow and the subsequent assessment of the degree of mixing under various flow and injection conditions.

Fig. 4.1: Cavity Transfer Mixer.

The CTM consists of a closely fitting rotor and stator, both of which are covered in staggered rows of hemispherical cavities to create a continuous helical path in which the fluid elements undergo a series of cutting, folding and turning deformations. This leads to an incease in distributive mixing along the length of the CTM. The narrow flow domain between the rotor and the stator gives rise to a region of high shear and elongational flow and it is precisely this feature of the CTM which could be modified to increase the dispersive mixing.

Fig 4.2: Pictures of CTM rotor (above left), the casing (above right) and our transparent cased CTM (below).

Fig. 4.3: The Transparent CTM case shows clearly the position of the cavites on both the Rotor (metal, rotates) and the Stator (transparent here, stationary).

4.9 Polymer Viscoelasticity

Viscoelastic techniques characterize the relative importance of elastic (solid-like) and viscous (liquid-like) response components in describing polymer behavior. The deformation response of polymers to applied stress or load is defined, and the architecture of the polymer chains can be determined.

Applications

- Molecular weight/branching
- Cure kinetics
- Energy dissipation/damping
- Blend compatibility
- Gelation and fusion of plastisols
- Stability
- Phase changes
- Shrinkage/dimensional stability
- Stress growth and relaxation
- Melt viscosity
- Dynamic mechanical properties (Tg, Tm, structure, etc.)
- Service life prediction (Time-Temperature Super-position)
- Compressive Creap/Creap Recovery

Polymer Rheology-I

Semester	*Methods of Education*						*Credits*	
	Lecture	*Recit*	*Lab.*	*Home work*	*Other*	*Total*	*Credit*	*ECTS Credit*
1	42	—	—	20	125.5	187.5	3	7.5

Language	Turkish
Compulsory/Elective	Optional
Prerequisites	—
Course contents	Rheology: Definition, properties of Newtonian and non-Newtonian fluids, viscosity, viscoelasticity, extentional viscosity, mechanical properties of polymers and their characterization, colloidal systems, rheology of polymeric liquids, rheology of suspensions, rheology and morphology.
Course Objectives	To learn the basic definitions of rheology, and the rheology of colloidal systems, suspensions, polymeric systems, jels and conducting polymers also the relationship between rheology and morphology.
Learning Outcomes and Competences	Basic principles of rheology and its applications to the real ssystems especially to polymers and suspensions will be discussed.
Textbook and/or References	An introduction to rheology, H.A. Barnes, J.F. Hutton, Elsevier, NY, 1989 Rheology of polymeric systems, P.J. carreau, D.C. De Kee, NY, 1997 Polymer Rheology, R.S. Lenk, Applied Sci. Publishers, London, 1978 4 Rheology for chemists, J.W. Goodwin, R.W. Hughes, RSC, NY, 2000

Assessment Criteria		*If any, mark as (X)*	*Percentage*
	Midterm Exams	X	40
	Quizzes	—	—
	Homeworks	X	10
	Projects	—	—
	Term Paper	—	—
	Laboratory Work	—	—
	Other	—	—
	Final Exam	X	50
Instructors	Prof. Dr. Halil ? brahim ÜNAL		

(*Contd.*)

Semester	*Methods of Education*						*Credits*	
	Lecture	*Recit*	*Lab.*	*Home work*	*Other*	*Total*	*Credit*	*ECTS Credit*

Week	*Subjet*
1	Rheology: definition, Properties of Newtonian and non-Newtonian fluids, definition of rheological parameters.
2	Rheology: definition, Properties of Newtonian and non-Newtonian fluids, definition of rheological parameters.
3	Viscosity: Factors that effects viscosity of fluids, shear thinning and shear thickinning fluids.
4	Investigation of thixothiropic behavior, types of viscometers used Viscoelasticity: extentional viscosity showing systems.
5	Investigation of mechanical properties of polymers and their characterisation
6	Midterm exam
7	Colloidal systems
8	Rheology of polymeric fluids
9	Properties of gels and their preparation, properties of liquid crystals
10	Rheology of suspensions
11	Rheology and morphology
12	Rheology of powder injection molding

5 Polymer Crystallization

Simple polymers typically form thin lamellar crystals, where the polymer chains are oriented approximately perpendicular to the plane of the lamella, folding back upon themselves at each surface. However, this morphology is not the most stable—the polymer crystal would have a lower energy if the polymers were completely extended with no folds. Therefore, the morphology must be a result of the dynamics of crystal growth. The two dominant theories (surface nucleation and Sadler-Gilmer) that attempt to explain this behaviour, particularly the dependence of the thickness and the growth rates of the crystal on temperature, make a number of assumptions about the microscopic mechanisms of crystallization. It was the aim of my post-doctoral research with Daan Frenkel to use computer simulation and modelling to test some of these assumptions.

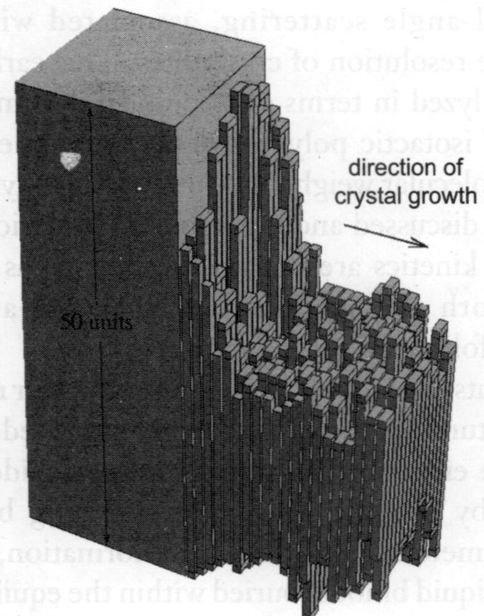

However, our simulations revealed a mechanism of thickness selection that differed from both the existing theories. We found that at a particular temperature there is only one dynamically-stable thickness at which crystals can grow with constant thickness. Thicker (thinner) crystals thin (thicken) as they grow until this critical thickness is reached. This mechanism is illustrated by the adjacent cut through a polymer crystal that was grown in our simulations. The crystal thins to the stable thickness within 5-10 layers and then continues to grow at that thickness. One way to provide experimental support for this mechanism would be to examine the profile of the steps on the surface of a polymer crystal that result from changes in temperature during growth.

5.1 Polymer Crystallization

Polymer crystallization is a very old problem. New experiments on the early stages of polymer crystallization reveal a precursor liquid-liquid phase separation in many polymers. This raises fascinating questions about metastability and crystallization in a wide variety of sytems, not just polymers!

Papers

An experimental study of the early stages of crystallization in iPP has shown a qualitative difference between the behavior at low supercooling with that observed with a deep quench. To address previous misgivings in the limits of resolution of crystallites by wide-angle scattering, a new detector has been used that has many orders of magnitude improvements in count rate. At low degrees of undercooling there is a substantial gap between the appearance of a peak in the small-angle scattering, associated with electron density modulations, and the resolution of crystallites. This early growth in electron density has been analyzed in terms of a spinodal decomposition process and the stability limit of isotactic polypropylene determined for three different samples of varying molecular weight. The underlying physics of the early stages of crystallization are discussed and a number of scenarios eliminated; at high temperature Avrami kinetics are not observed whereas at low temperatures the structure in both the small-angle and wide-angle regimes grow contemporaneously following secondary nucleation.

Recent experiments have shown that in some polymer melts quenched below the melting temperature, spinodal kinetics are observed in small-angle X ray scattering before the emergence of Bragg peaks at wide angles. We explain these observations by proposing that the coupling between density and secondary order parameters (chiefly chain conformation, but also orientation) gives rise to a liquid-liquid binodal buried within the equilibrium liquid-crystal coexistence region. A simple phenomenological theory is developed to illustrate the idea. Shear is shown to enhance the kinetic role of this hidden binodal. A number of experimentally testable consequences of this model are discussed, including the effects of the 'hidden' critical point on crystallization kinetics.

5.2 Insights into Early Stages of Polymer Crystallization

Because of their diverse physical and mechanical properties, polymers—such as polyethylene—have become essential materials in a variety of defense systems as well as medical, industrial, and environmental applications. Typically, polymers are composed of both crystalline and amorphous regions, and their properties are strongly influenced by the size and shape of the crystalline regions. Understanding how and why these regions form will help us control these properties, leading to new applications and improvements to existing materials. However, the atomistic processes that lead to polymer crystallization are not well formulated, and experiments have not given us the level of detail we need. Using LLNL's terascale computational capabilities, we are able to further our understanding of these processes on an atomistic level.

Relevance to CMELS Research Themes

By increasing our understanding of polymer crystallization, we are gaining fundamental knowledge about the nature of crystal formation and, more specifically, the aging of technologically important materials in a variety of defense systems.

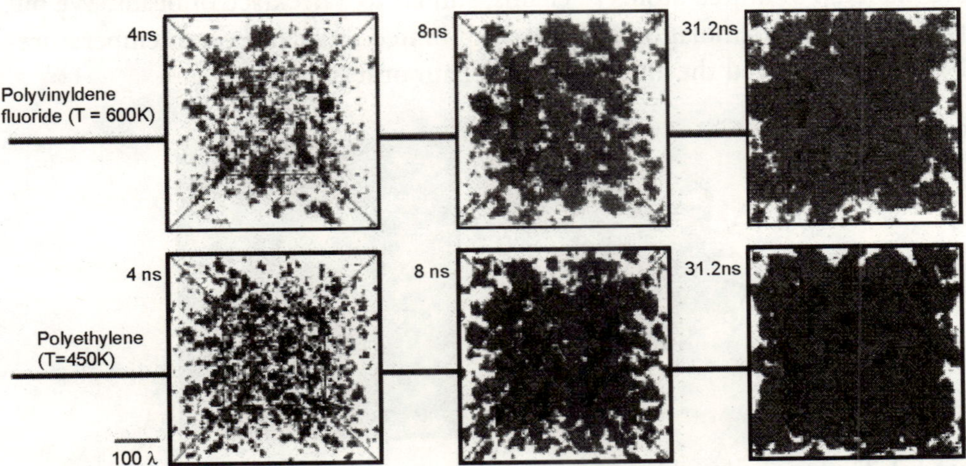

Fig. 5.1: In the largest molecular dynamics simulations of polymer crystallization to date, our models show the evolution of the spinodal-assisted crystallization process for the polar (orange atoms) and nonpolar (grey atoms) polymers. The crystalline regions are explicitly shown, while the amorphous domains are shown as the white space in each panel.

The polymer materials in defense systems such as nuclear weapons must perform within predictable margins, and we are turning increasingly to simulations to ensure that these materials will perform as required. In our polymer crystallization investigations, we have focused on examining high-explosive aging phenomena relevant to the Stockpile Stewardship Program's Enhanced Surveillance Campaign. Our goal is to obtain, through accurate models, a lifetime assessment of physical and mechanical properties of polymers relevant to their use in national defense systems. In a broader sense, the results of our work can also be applied in atomic-scale control of the physical and mechanical properties of polymers.

Major Accomplishments in 2005

Under some circumstances, polymers nucleate almost instantly, in contrast to classical nucleation, where a crystal seed is formed over time. Using the LLNL 23-TFLOP/s Thunder supercomputer and other computational resources, we conducted the largest molecular dynamics simulations of polymer crystallization to date to test recent experimental findings and theoretical predictions that, when rapidly cooled, polymeric materials undergo instant crystallization (known as spinodally assisted nucleation). Understanding spinodally assisted nucleation at the atomistic level is the first step to elucidating the pretransition state of polymer crystallization.

We considered two classes of polymers: polar polymers (polyvinylidene fluoride, or PVDF) and nonpolar polymers (polyethylene, or PE). Both classes are made up of united atoms in chains, similar to a necklace of beads. We put these chains in a simulation cell and equilibrated them at several temperatures before we observed the almost instant chain ordering.

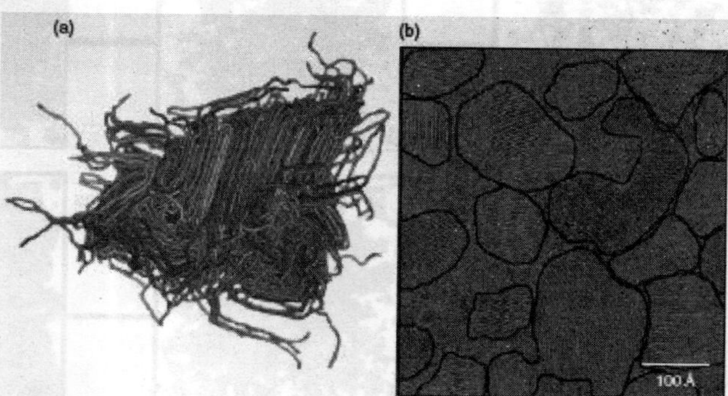

Fig. 5.2: The first panel (a) shows a single ordered polymer domain with a morphology for the polar polymer model. Individual polymers are colored to illustrate polymer chains entering the amorphous region from an ordered region and adjoining two separate adjacent ordered domains (adjacent domains are not shown). The second panel (b) shows a representative ensemble of ordered domains.

To address questions about the atomistic mechanism responsible for this instant ordering, our simulations used the length scale of 10 to 20 nm—the range at which spinodally assisted nucleation occurs. Using simulation cells twice the size of our length scale, our results provided atomistic information of polymer nucleation with unprecedented resolutions. (For example, the number of beads in one nonpolar melt was five million, and the size of the simulation cell was approximately 50 nm.)

Confirming earlier experimental findings, our simulations revealed microphase separation of bulk amorphous polymer ensembles into many ordered domains (Figure 5.1). These domains coalesce and grow, initially forming a small number of crystalline regions that grow to occupy most of the simulation volume at the end of the runs. Figure 5.2(a) shows an enlarged ordered domain isolated from the sample, which reveals an interface between oriented and unoriented domains. Ordered crystalline domains are circled in Figure 5.2(b) to demonstrate the complex nature of the polymer morphology.

Scientific Impact

As the first atomistic observation of the debated prenucleation, this work provides unambiguous insight into the physics of the early stages of polymer ordering leading to crystallization. Furthermore, with the unprecedented resolutions and details of our simulations, it was possible to compare and confirm our findings with laboratory experiments. Our results pertaining to crystallite size and structure as well as the critical length of polymer segments provide crucial information needed to form a fundamental understanding of the structure–property relationships of semicrystalline polymers.

New Frontiers

We will use our findings about bulk polymer melts to give us greater control of the crystallite microstructure, orientation, density, and size—all of which contribute to performance. Not only can we find specific defense applications in the case of binding materials for high explosives, we also expect broader applications such as polymer transistors, which offer great flexibility at low cost. We are looking at ways to control the microstructure during manufacturing, when semiconductor films are deposited by various techniques (spin coating, jet printing, dip coating) on a dielectric surface.

Furthermore, our findings about spinodally assisted nucleation are applicable to other materials (such as metals), and we are using this process to develop new materials for National Ignition Facility targets and other applications. Powerful simulations on our terascale supercomputers will help us understand and control the microstructures of polymers and metals, leading to new and improved materials in a broad range of technologically and scientifically important devices.

5.3 Outline

1. General background,
2. Computational method,
3. Some recent results,
4. Current research projects,
5. Collaboration, and
6. Related papers.

1. General Background

Many polymers, whose properties favor the formation of locally ordered structures, crystallize at low temperatures. However, the connectivity of the long chain molecules usually hinders the formation of perfect crystals and leads to semicrystalline materials. For these heterogeneous materials one would like to better understand the pathways leading to a particular composition of crystalline and amorphous regions as well as the influence of external control parameters on the resulting structures.

2. Computational Method

We try to tackle this problem by direct molecular dynamics simulations of a mesoscopic coarse-grained model for poly-vinylalcohol (CG-PVA). Our model goes one step further than usual united-atom models and resumes all atoms of a monomer into one sphere, and it contains a specific angular bending potential. This approach appeares to be extremely efficient.

3. Some Recent Results

With these approximations, the crystallization from the homogeneous nuclueation in the melt up to the formation of chain-folded crystal lamellae can be observed. Furthermore, it yields insight into the nucleation process, the growth of the crystal and into the single chain conformations in the crystal. See the figure.

We also discuss the influence of different simulation parameters, and in particular of the angular potential. We characterized the melt at a reference temperature and looked for correlations with the crystallization temperature determined during continuous cooling. For most quantities as persistence length, radius of gyration or relaxation times, no trivial correlation can be found, except for the fraction of stretched *tt* conformations in the melt: the higher this fraction, the easier is the crystallization and thus the higher the temperature where ordering starts.

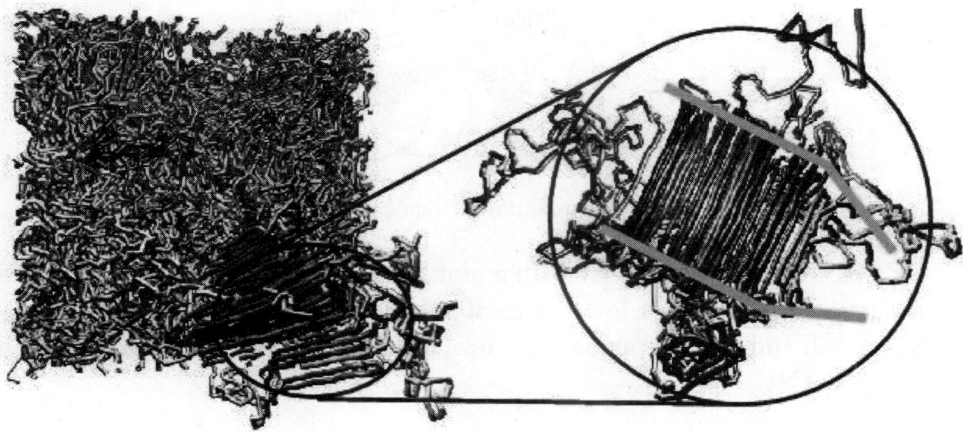

Fig. 5.3: Formation of a crystal lamella as observed with molecular dynamics. The simulation box contains 500 monodisperse chains of length N=400. Right: Visualization of 13 chains of the first crystalline domaine. The crystallization domain is observed to extend over several chain diameters.

4. Current research Projects

In near future we plan to study the following topics:

- Further improvement of coarse-grained models for chemical polymers,
- Study of single chain properties (collapse with decreasing temperature, etc.),
- Influence of chain length, cooling rate, etc. on the crystallization process and the resulting structures, and
- Structure formation of crystallizable polymer melts close to solid, structured substrates. Due to the large loss of conformational entropy the crystallization process, especially at weak undercooling, is often triggered by container walls, dispersed foreign particles and other heterogeneities. Our numerical studies (H. Meyer) are accompanied by theoretical developments (A. Johner).

5. Collaboration

Some polymers form more crystalline solids than others. It will be useful for us to relate the tendency to crystallize to the chemical composition and structural details of particular polymers.

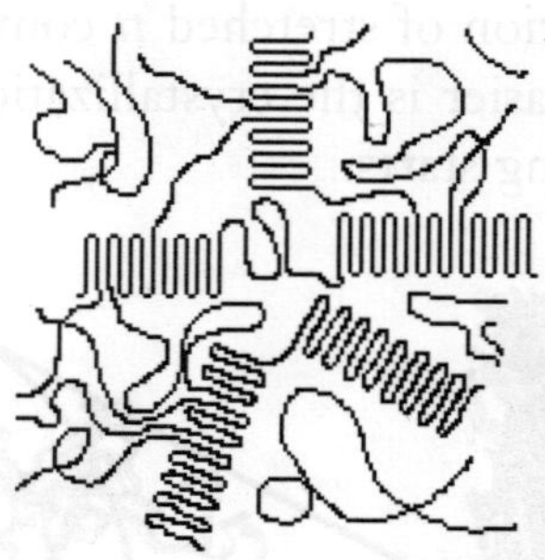

Crystallization Tendency

Six factors favor a polymer with a high percent crystallinity: a regular and symmetrical linear chain, a low degree of polymerization, strong intermolecular forces, small and regular pendant groups, a slow rate of cooling, and oriented molecules.

Structural Regularity

- ❖ Degree of Polymerization
- ❖ Intermolecular Forces
- ❖ Pendant Groups
- ❖ Processing:
  - Rate of Cooling, and
  - Orientation.

Structural Regularity

To crystallize a polymer chain must be linear, although limited crystallization can take place if a small number of branches are present. Crystallization is favored by a regular arrangement along the polymer chain giving the structure a high degree of symmetry. Linear polyethylene for example can form a solid with over 90 per cent crystallinity in some cases. This is made possible by the planar zig-zag structure easily assumed by the molecule.

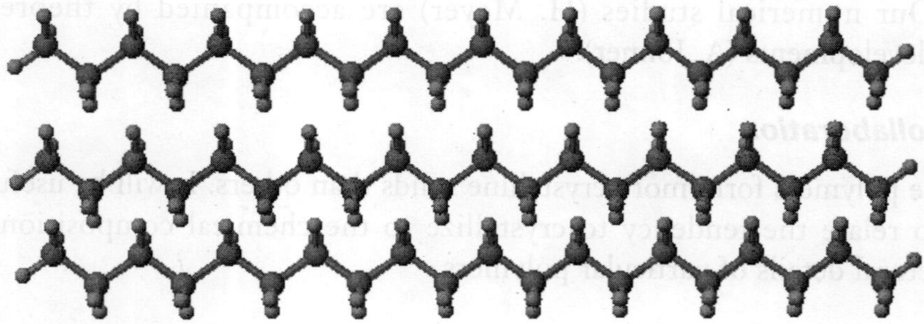

Normal polystyrene is atactic with no regular order in the position of the benzene rings along the chain. The irregularity prevents the chains from packing closely to each other. Atactic polystyrene, is amorphous. It is comparatively soft, low melting, and becomes swollen in solvents.

$—CH_2—CH—CH_2—CH—CH_2—CH—CH_2—CH—CH_2—CH—CH_2—CH—$

Atactic Polystyrene

In syndiotactic polystyrene the benzene rings are on alternate sides of the chain. This allows the chains to pack into crystals.

Syndiotactic polystyrene is crystalline. It is rigid, high melting, and not penetrated readily by solvents.

$—CH_2—CH—CH_2—CH—CH_2—CH—CH_2—CH—CH_2—CH—CH_2—CH—$

$—CH_2—CH—CH_2—CH—CH_2—CH—CH_2—CH—CH_2—CH—CH_2—CH—$

Syndiotactic Polystyrene (two chains shown)

Degree of Polymerization

Relatively short polymer chains form crystals more readily than long chains, because the long chains tend to be more tangled[27]. High crystallinity generally means a stronger material, but low molecular weight polymers usually are weaker in strength even if they are highly crystalline. Low molecular weight polymers have a low degree of chain entanglement, so the polymer chains can slide by each other and cause a break in the material.

Intermolecular Forces

Crystallinity is favored by strong interchain forces. The presence of polar and hydrogen bonding groups favors crystallinity because they make possible dipole-dipole and hydrogen bonding intermolecular forces. A polyester, such as poly(ethylene terephalate), contains polar ester groups. Dipole-dipole forces between the polar groups hold the PET molecules in strong crystals.

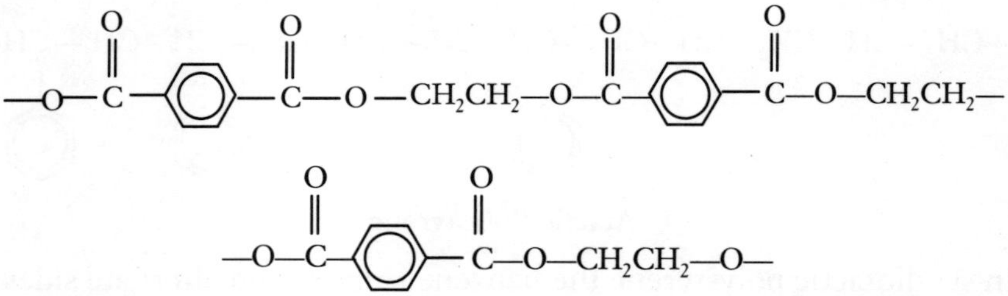

$$-O-\overset{O}{\overset{\|}{C}}-C_6H_4-\overset{O}{\overset{\|}{C}}-O-CH_2CH_2-O-\overset{O}{\overset{\|}{C}}-C_6H_4-\overset{O}{\overset{\|}{C}}-O-CH_2CH_2-$$

$$-O-\overset{O}{\overset{\|}{C}}-C_6H_4-\overset{O}{\overset{\|}{C}}-O-CH_2CH_2-O-$$

Crystallinity in poly(ethylene terephalate) also is favored by the structural regularity of the benzene rings in the chain. The benzene rings stack together in an orderly fashion.

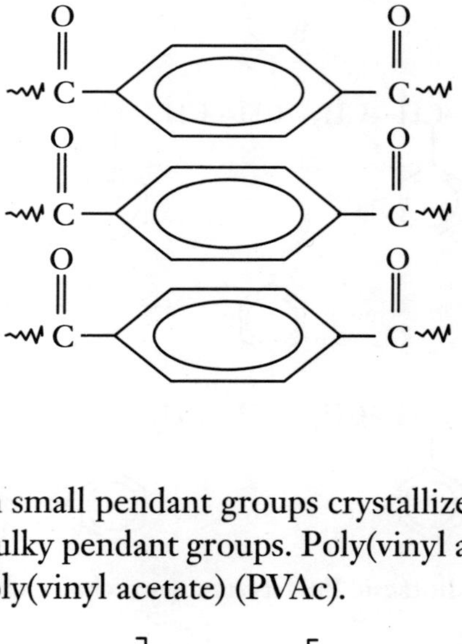

$$\sim\overset{O}{\overset{\|}{C}}-C_6H_4-\overset{O}{\overset{\|}{C}}\sim$$
$$\sim\overset{O}{\overset{\|}{C}}-C_6H_4-\overset{O}{\overset{\|}{C}}\sim$$
$$\sim\overset{O}{\overset{\|}{C}}-C_6H_4-\overset{O}{\overset{\|}{C}}\sim$$

Pendant Groups

Regular polymers with small pendant groups crystallize more readily than do polymers with large, bulky pendant groups. Poly(vinyl alcohol) (PVA) is made by the hydrolysis of poly(vinyl acetate) (PVAc).

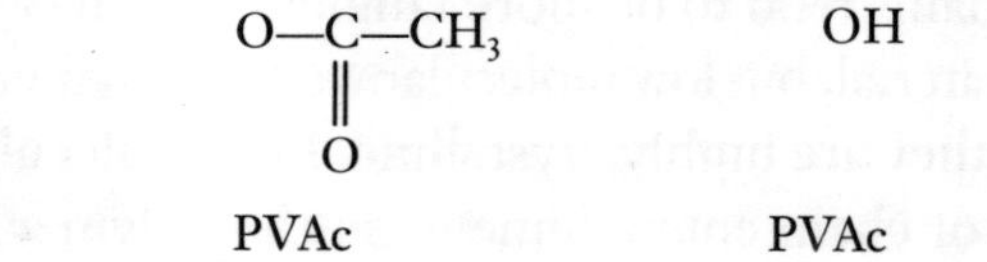

$$\left[-CH_2-\underset{\underset{\underset{O}{\|}}{O-C-CH_3}}{CH}-\right]_n \longrightarrow \left[-CH_2-\underset{OH}{CH}-\right]_n$$

PVAc PVAc

PVA crystallizes more readily than PVAc because of the bulky acetate groups in PVAc. The -OH groups in PVA also form strong hydrogen bonds.

Processing

A major difference between small molecules and polymers is that the morphology of a polymer is dependent on its thermal history. The crystallinity of a polymer can be changed by cooling the polymer melt slowly or quickly, and by "pulling" the bulk material either during its synthesis or during its processing

A. Cooling Rate

When they are processed industrially, polymers often are cooled rapidly from the melt. In this situation, crystallization is controlled by kinetics rather than thermodynamics. There may not be time for the chains, which are entangled in the melt, to separate enough to form crystals, so the amorphous nature of the melt is "frozen into" the solid. A polymer is more likely to have a higher percent crystallinity if it is cooled slowly from the melt.

B. Orientation

Crystallinity can be enhanced by pulling the bulk material either when it is synthesized or during its processing. This is common for both films and fibers.

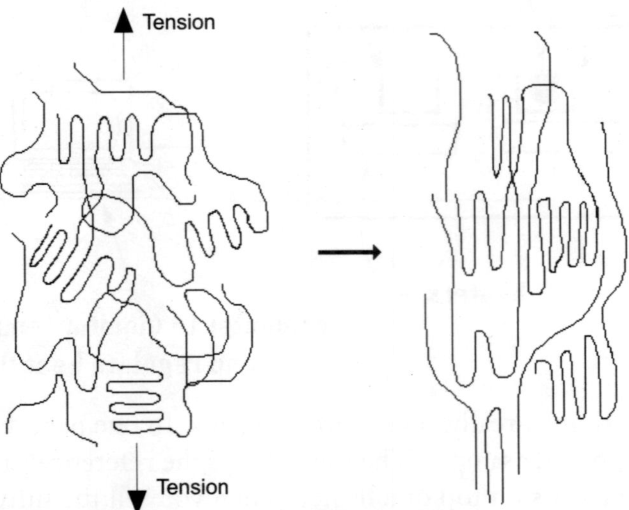

When a film is formed the small crystallites tend to be randomly oriented relative to each other. Drawing (stretching) the film pulls the individual chains into a roughly parallel organization as is shown in the schematic diagram at the right. Films can either be uniaxially oriented (oriented in only one direction)

or biaxially oriented (oriented in two directions). Fibers normally are drawn so that they are oriented in one direction. Unstretched nylon fibers are brittle, for example. When the fibers are stretched the oriented fibers are strong and tough. Polyethylene can be unentangled by forming a gel with a low molecular weight solvent. When the gel is drawn, the resulting fibers are highly oriented. Ultra-oriented PE formed in this way is used in bullet-proof vests.

5.4 Differential Scanning Calorimetry

Differential scanning calorimetry is a technique we use to study what happens to polymers when they're heated. We use it to study what we call the *thermal transitions* of a polymer. And what are thermal transitions? They're the changes that take place in a polymer when you heat it. The melting of a crystalline polymer is one example. The glass transition is also a thermal transition.

So how do we study what happens to a polymer when we heat it? The first step would be to heat it, obviously. And that's what we do in differential scanning calorimetry, or DSC for short.

We heat our polymer in a device that looks something like this:

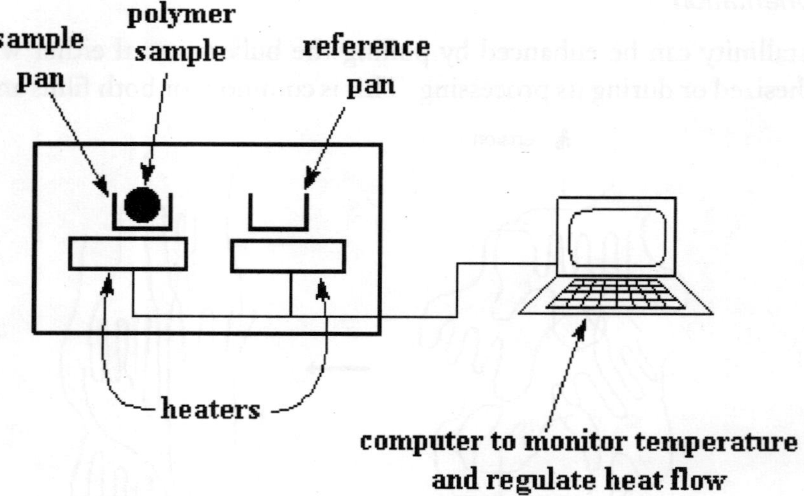

It's pretty simple, really. There are two pans. In one pan, the sample pan, you put your polymer sample. The other one is the reference pan. You leave it empty. Each pan sits on top of a heater. Then you tell the nifty computer to turn on the heaters. So the computer turns on the heaters, and tells it to heat the two pans at a specific rate, usually something like 10 °C per minute. The computer makes absolutely sure that the heating the rate stays exactly the same throughout the experiment.

But more importantly, it makes sure that the two separate pans, with their two separate heaters, heat at the same rate as each other.

Huh? Why wouldn't they heat at the same rate? The simple reason is that the two pans are different. One has polymer in it, and one doesn't. The polymer sample means there is extra material in the sample pan. Having extra material means that it will take more heat to keep the temperature of the sample pan increasing at the same rate as the reference pan.

So the heater underneath the sample pan has to work harder than the heater underneath the reference pan. It has to put out more heat. By measuring just *how much* more heat it has to put out is what we measure in a DSC experiment.

Specifically what we do is this: We make a plot as the temperature increases. On the *x*-axis we plot the temperature. On the *y*-axis we plot difference in heat output of the two heaters at a given temperature.

Heat Capacity

We can learn a lot from this plot. Let's imagine we're heating a polymer. When we start heating our two pans, the computer will plot the difference in heat output of the two heaters against temperature. That is to say, we're plotting the heat absorbed by the polymer against temperature. The plot will look something like this at first.

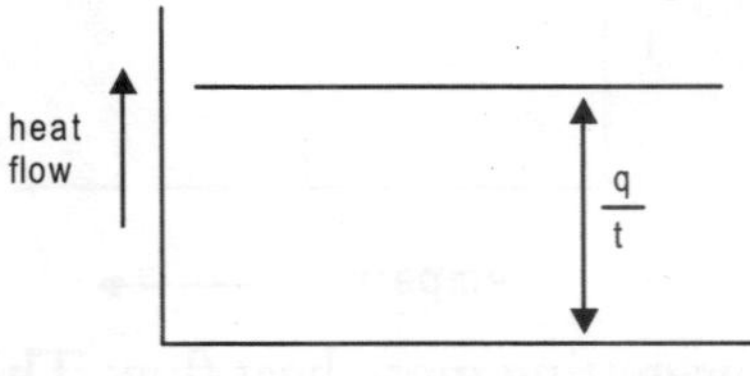

The heat flow at a given temperature can tell us something. The heat flow is going to be shown in units of heat, *q* supplied per unit time, *t*. The heating rate is temperature increase *T* per unit time, *t*. Got it?

$$\frac{heat}{time} = \frac{q}{t} = heat\ flow$$

$$\frac{Temperature\ increase}{time} = \frac{\Delta T}{t} = heat\ rate$$

Let's say now that we divide the heat flow *q/t* by the heating rate *T/t*. We end up with heat supplied, divided by the temperature increase.

$$\frac{\frac{q}{t}}{\frac{\Delta T}{t}} = \frac{q}{\Delta T} = Cp = \textit{heat capacity}$$

Remember from the glass transition page that when you put a certain amount of heat into something, its temperature will go up by a certain amount, and the amount of heat it takes to get a certain temperature increase is called the *heat capacity*, or C_p. We get the heat capacity by dividing the heat supplied by the resulting temperature increase. And that's just what we've done in that equation up there. We've figured up the heat capacity from the DSC plot.

The Glass Transition Temperature

Of course, we can learn a lot more than just a polymer's heat capacity with DSC. Let's see what happens when we heat the polymer a little more. After a certain temperature, our plot will shift upward suddenly, like this:

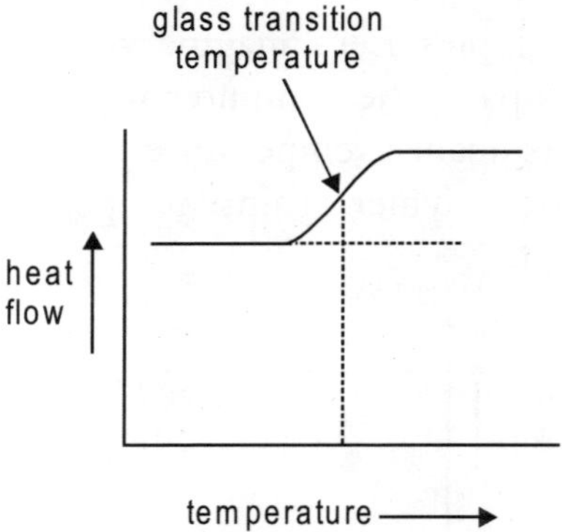

This means we're now getting more heat flow. This means we've also got an increase in the heat capacity of our polymer. This happens because the polymer has just gone through the glass transition. And as you learned on the glass transition page, polymers have a higher heat capacity above the glass transition temperature than they do below it. Because of this change in heat capacity that occurs at the glass transition, we can use DSC to measure a polymer's glass transition temperature. You may notice that the change doesn't occur suddenly, but takes place over a temperature range. This makes picking one discreet *Tg* kind of tricky, but we usually just take the middle of the incline to be the *Tg*.

Crystallization

But wait there is more, so much more. Above the glass transition, the polymers have a lot of mobility. They wiggle and squirm, and never stay in one position for very long. They're kind of like passengers trying to get comfortable in airline seats, and never quite succeeding, because they can move around more. When they reach the right temperature, they will have gained enough energy to move into very ordered arrangements, which we call crystals, of course.

When polymers fall into these crystalline arrangements, they give off heat. When this heat is dumped out, it makes the little computer-controlled heater under the sample pan really happy. It's happy because it doesn't have to put out much heat to keep the temperature of the sample pan rising. You can see this drop in the heat flow as a big dip in the plot of heat flow versus temperature:

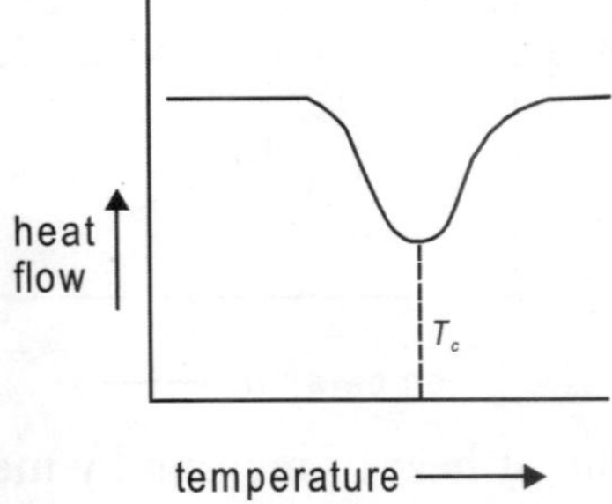

This dip tells us a lot of things. The temperature at the lowest point of the dip is usually considered to be the polymer's crystallization temperature, or T_c. Also, we can measure the area of the dip, and that will tell us the latent energy of crystallization for the polymer. But most importantly, this dip tells us that the polymer can in fact crystallize. If you analyzed a 100 per cent amorphous polymer, like atactic polystyrene, you wouldn't get one of these dips, because such materials don't crystallize.

Also, because the polymer gives off heat when it crystallizes, we call crystallization an *exothermic* transition.

Melting

Heat may allow crystals to form in a polymer, but too much of it can be their undoing. If we keep heating our polymer past its T_c, eventually we'll reach another thermal transition, one called melting. When we reach the polymer's melting temperature, or T_m, those polymer crystals begin to fall apart, that is they melt. The chains come out of their ordered arrangements, and begin to move around freely. And in case you were wondering, we can spot this happening on a DSC plot.

Remember that heat that the polymer gave off when it crystallized? Well when we reach the T_m, it's payback time. There is a latent heat of melting as well as a latent heat of crystallization. When the polymer crystals melt, they must absorb heat in order to do so. Remember melting is a first order transition.

This means that when you reach the melting temperature, the polymer's temperature won't rise until all the crystals have melted. This means that the little heater under the sample pan is going to have to put a lot of heat into the polymer in order to both melt the crystals and keep the temperature rising at the same rate as that of the reference pan. This extra heat flow during melting shows up as a big peak on our DSC plot, like this:

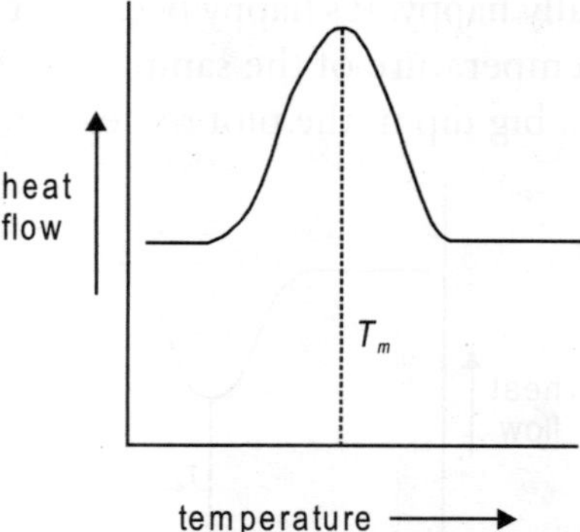

We can measure the latent heat of melting by measuring the area of this peak. And of course, we usually take the temperature at the top of the peak to be the polymer's melting temperature, T_m. Because we have to add energy to the polymer to make it melt, we call melting an *endothermic* transition.

Putting It All Together

So let's review now: we saw a step in the plot when the polymer was heated past its glass transition temperature. Then we saw a big dip when the polymer reached its crystallization temperature. Then finally we saw a big peak when the polymer reached its melting temperature. To put them all together, a whole plot will often look something like diagram on next page.

Of course, not everything you see here will be on every DSC plot. The crystallization dip and the melting peak will only show up for polymers that can form crystals. Completely amorphous polymers won't show any crystallization, or any melting either. But polymers with both crystalline and amorphous domains, will show all the features you see above.

If you look at the DSC plot you can see a big difference between the glass transition and the other two thermal transitions, crystallization and melting. For the glass transition, there is no dip, and there's no peak, either. This is because there is no latent heat given off, or absorbed, by the polymer during the glass transition. Both melting and crystallization involve giving off or absorbing heat. The only thing we do see at the glass transition temperature is a change in the heat capacity of the polymer.

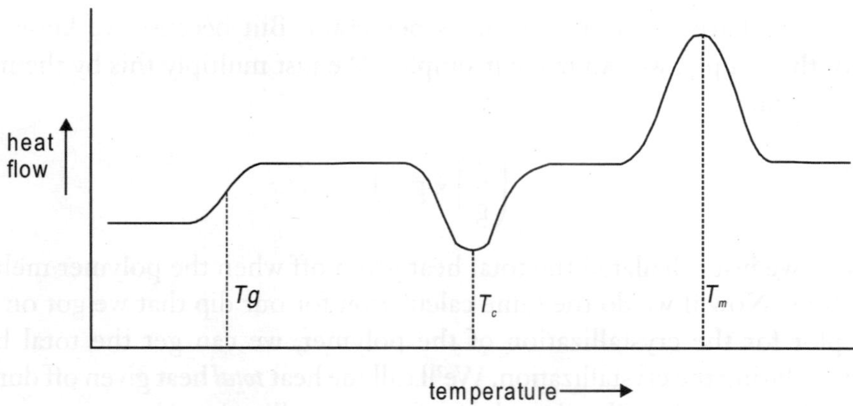

The entire DSC plot, right before your very eyes!

Because there is a change in heat capacity, but there is no latent heat involved with the glass transition, we call the glass transition a *second order transition*. Transitions like melting and crystallization, which do have latent heats, are called *first order transitions*.

How Much Crystallinity

DSC can also tell us how much of a polymer is crystalline and how much is amorphous. If you read the page dealing with polymer crystallinity, you know that many polymers contain both amorphous and crystalline material. But how much of each? DSC can tell us. If we know the latent heat of melting, ΔH_m, we can figure out the answer.

The first thing we have to do is measure the area of that big peak we have for the melting of the polymer. Now our plot is a plot of heat flow per gram of material, versus temperature. Heat flow is heat given off per second, so the area of the peak is given is units of heat x temperature × time^{-1} × mass^{-1}. We usually would put this in units such as joules × kelvins × (seconds)$^{-1}$ x (grams)$^{-1}$:

$$\text{area} = \frac{\text{heat} \times \text{temperature}}{\text{time} \times \text{mass}} = \frac{\text{JK}}{\text{sg}}$$

Got that? Don't worry. It gets simpler. We usually divide the area by the heating rate of our dsc experiment. The heating rate is in units of K/s. So the expression becomes simpler:

$$\frac{\text{area}}{\text{heating rate}} = \frac{\frac{\text{JK}}{\text{sg}}}{\frac{\text{K}}{\text{s}}} = \frac{\text{J}}{\text{g}}$$

Now we have a number of joules per gram. But because we know the mass of the sample, we can make it simpler. We just multiply this by the mass of the sample:

$$\left(\frac{J}{g}\right) \times g = J$$

Now we just calculated the total heat given off when the polymer melted. Neat, huh? Now if we do the same calculation for our dip that we got on the DSC plot for the crystallization of the polymer, we can get the total heat absorbed during the crystallization. We'll call the heat *total* heat given off during melting $H_{m, total}$, and we'll call the heat of the crystallization $H_{c, total}$.

Now we're going to subtract the two:

$$H_{m, total} - H_{C, total} = H'$$

Why did we just do that? And what does that number *H*' mean? *H*' is the heat given off by that part of the polymer sample which was already in the crystalline state *before* we heated the polymer above the T_c. We want to know how much of the polymer was crystalline before we induced more of it to become crystalline. That's why we subtract the heat given off at crystallization. Is everyone following me?

Now with our magic number *H*' we can figure up the percent crystallinity. We're going to divide it by the specific heat of melting, H_c^*. The specific heat of melting? That's the amount of heat given off by a certain amount, usually one gram, of a polymer. *H*' is in joules, and the specific heat of melting is usually given in joules per gram, so we're going to get an answer in grams, which we'll call m_c.

$$\frac{H'}{H^2 m} = m_c \qquad \frac{J}{\frac{J}{g}} = g$$

This is the total amount of grams of polymer that were crystalline below the T_c. Now if we divide this number by the weight of our sample, m_{total}, we get the fraction of the sample that was crystalline, and then of course, the percent crystallinity:

$$\frac{m_c}{m_{total}} = \textit{crystalline fraction}$$

$$\text{crystalline fraction} \times 100 = \%\ \text{crystallinity}$$

And that's how we use DSC to get percent crystallinity.

5.5 Polymer Crystalinity

We're talking about another kind of crystal here. The kind of crystal we're talking about here is any object in which the molecules are arranged in a regular order and pattern. Ice is a crystal. In ice all the water molecules are arranged in a specific manner. So is table salt, sodium chloride. (Oddly, your mother's good crystal drinking glasses are not crystal at all, as glass is an *amorphous solid*, that is, a solid in which the molecules have no order or arrangement.)

To understand all this talk of crystals and amorphous solids, it helps to go home. Go home? Why? So you can look in your sock drawer, that's why. You see, some people are very neat and orderly. When they put their socks away they fold them and stack them very neatly. Like this:

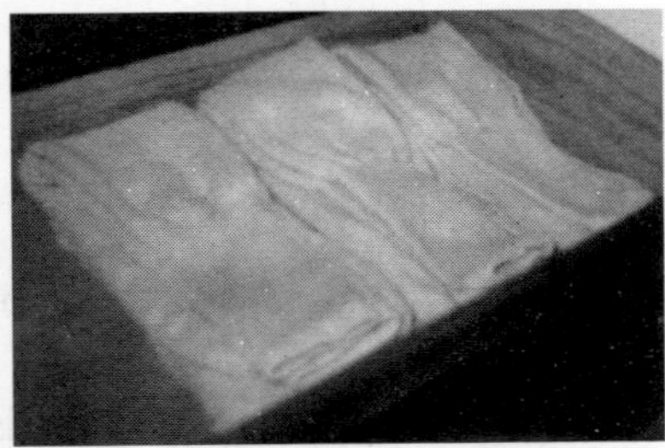

Other people don't really care about how neat their sock drawers look. Such folk will just throw their socks in the drawer in one big tangled mess. Their sock drawers look like this:

Polymers are just like socks in that sometimes they are arranged in a neat orderly manner, like the sock drawer in the top picture. When this is the case, we say the polymer is *crystalline*. Other times there is no order, and the polymer chains just form a big tangled mess, like the socks in the bottom picture. When this happens, we say the polymer is *amorphous*. We're going to talk about the neat and orderly crystalline polymers on this page. So what kind of arrangements do the polymers like to form?

They like to line up all stretched out, kind of like a neat pile of new boards down at the lumber yard. But they can't always stretch out that straight. In fact, very few polymers can stretch out perfectly straight, and those are ultra-high molecular weight polyethylene, and aramids like Kevlar and Nomex. Most polymers can only stretch out for a short distance before they fold back on themselves. You can see this in the picture.

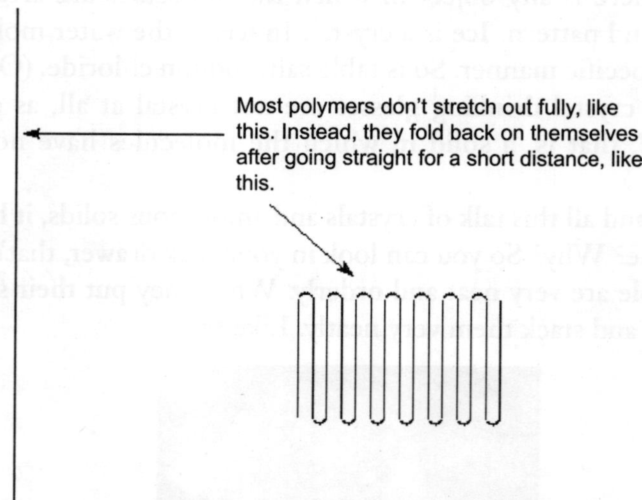

For polyethylene, the length the chains will stretch before they fold is about 100 angstroms.

But not only do polymers fold like this. Polymers also form stacks of these folded chains. There is a picture of a stack, called a lamella, right below:

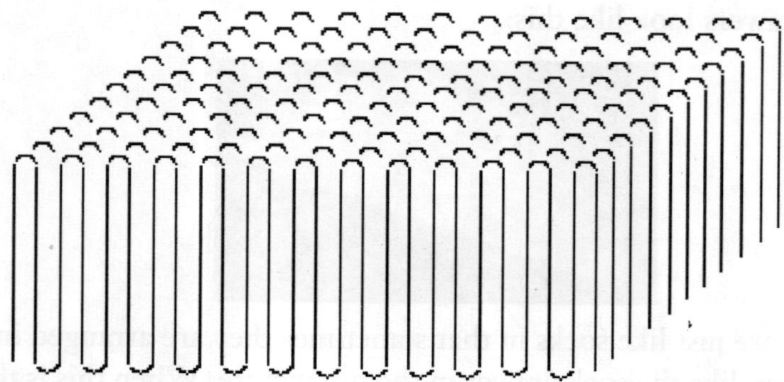

They can fold, and they can stack. A stack of polymer chains folded back on themselves like this is called a lamella.

Of course, it isn't always as neat as this. Sometimes part of a chain is included in this crystal, and part of it isn't. When this happens we get the kind of mess you see below. Our lamella is no longer neat and tidy, but sloppy, with chains hanging out of it everywhere!

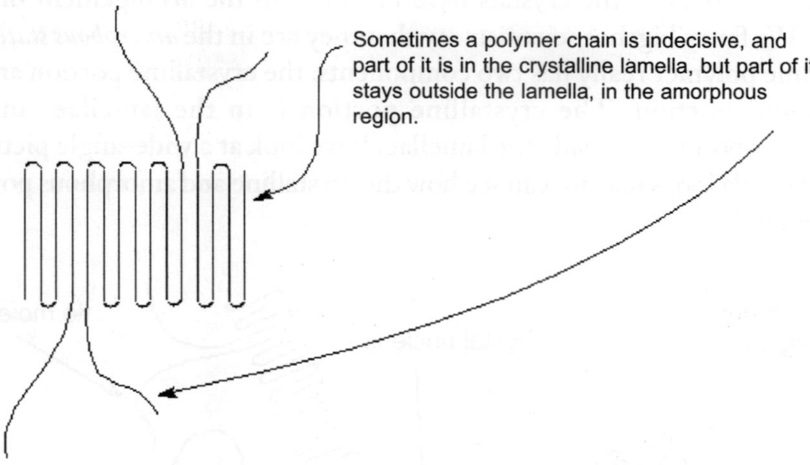

Of course, being indecisive, the polymer chains will often decide they want to come back into the lamella after wandering around outside for awhile. When this happens, we get a picture like this:

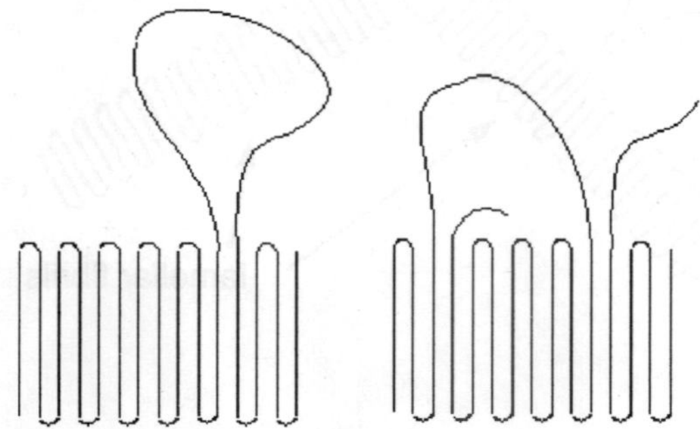

These lamellane have chains that go out for awhile, then come back in. On the left, the chain re-enters the lamellae right next to where it left. On the right our outgoing chain comes back in some distance away from where it left. Both are possible. These two pictures both show what is called the switchboard model of a polymer crystalline lamella.

This is the *switchboard model* of a polymer crystalline lamella. Because we like you, we're going to tell you that when a polymer chain doesn't wander around outside the crystal, but just folds right back in on itself, like we saw in the first pictures, that is called the *adjacent re-entry model.*

Amorphousness and Crystallinity

Are you wondering about something? If you look at those pictures up there, you can see that some of the polymer is crystalline, and some is not! Yes folks, most crystalline polymers are not entirely crystalline. The chains, or parts of chains, that aren't in the crystals have no order to the arrangement of their chains. We fancy bigshot scientists say that they are in the *amorphous state*. So a crystalline polymer really has two components: the crystalline portion and the amorphous portion. The crystalline portion is in the lamellae, and the amorphous potion is outside the lamellae. If we look at a wide-angle picture of what a lamella looks like, we can see how the crystalline and amorphous portions are arranged.

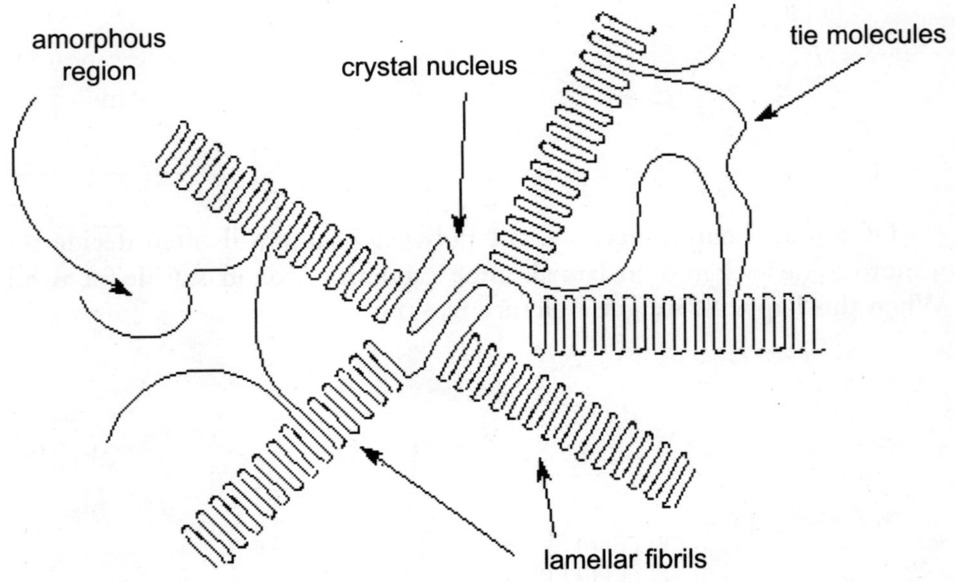

a polymer cyrstalline spherulite

As you can see, lamella grow like the spokes of a bicycle wheel from a central nucleus. (Sometimes we bigshot scientists like to call these spokes "lamellar fibrils".) The fibrils grow out in three dimensions, so they really look more like spheres than wheels. The whole assembly is called a *spherulite*. In a sample of a crystalline polymer weighing only a few grams, there are many billions of spherulites.

In between the crystalline lamellae, there are regions where there is no order to the arrangement of the polymer chains. These disordered regions are the amorphous regions we were talking about.

As you can also see in the picture, a single polymer chain may be partly in a crystalline lamella, and partly in the amorphous state. Some chains even start in one lamella, cross the amorphous region, and then join another lamella. These chains are called *tie molecules*.

So you see, no polymer is completely crystalline. If you're making plastics, this is a good thing. Crystallinity makes a material strong, but it also makes it brittle. A completely crystalline polymer would be too brittle to be used as plastic. The amorphous regions give a polymer *toughness*, that is, the ability to bend without breaking. But for making fibers, we like our polymers to be as crystalline as possible. This is because a fiber is really a long crystal. Want to know more? Then visit the Fiber Page! Many polymers are a mix of amorphous and crystalline regions, but some are highly crystalline and some are highly amorphous. Here are some of the polymers that tend toward the extremes:

Some Highly Crystalline Polymers:	*Some Highly Amorphous Polymers:*
Polypropylene	Poly(methyl methacrylate)
Syndiotactic polystyrene	Atactic polystyrene
Nylon	Polycarbonate
Kevlar and Nomex	Polyisoprene
Polyketones	Polybutadiene

So why is it that some polymers are highly crystalline and some are highly amorphous? There are two important factors, polymer structure and intermolecular forces.

Crystallinity and Polymer Structure

A polymer's structure affects crystallinity a good deal. If it's regular and orderly, it will pack into crystals easily. If it's not, it won't. It helps to look at polystyrene to understand how this works.

$$-\!\!\left[CH_2 - CH(C_6H_5) \right]_n\!\!-$$

polystyrene

As you can see on the lists above, there are two kinds of polystyrene. There is atactic polystyrene, and there is syndiotactic polystyrene. One is very crystalline, and one is very amorphous.

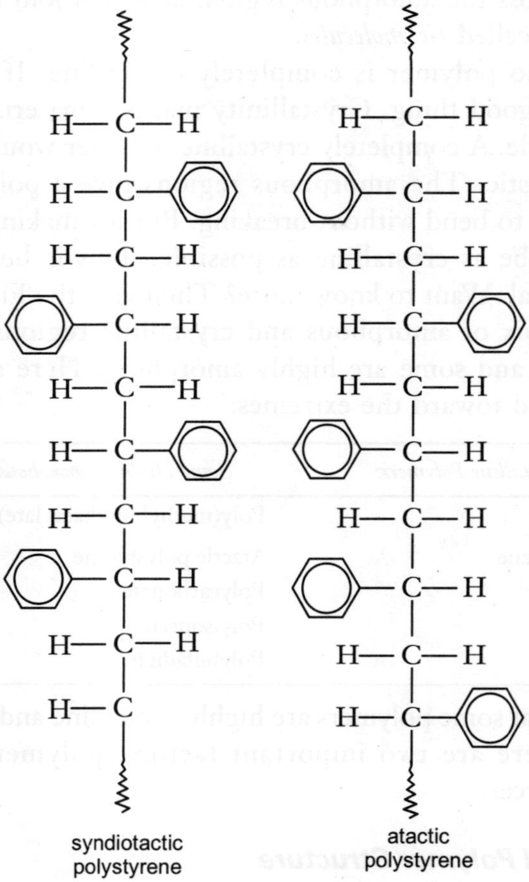

syndiotactic polystyrene

atactic polystyrene

Syndiotactic polystyrene has a regular structure, so it can pack into crystal structures. The irregular atactic polystyren can't.

Syndiotactic polystyrene is very orderly, with the phenyl groups falling on alternating sides of the chain. This means it can pack very easily into crystals.

But atactic styrene has no such order. The phenyl groups come on any which side of the chain they please. With no order, the chains can't pack very well. So atactic polystyrene is very amorphous.

Other atactic polymers like poly(methyl methacrylate) and poly(vinyl chloride) are also amorphous. And as you might expect, stereoregular polymers like isotactic polypropylene and polytetrafluoroethylene are highly crystalline.

Polyethylene is another good example. It can be crystalline or amorphous. Linear polyethylene is nearly 100 per cent crystalline. But the branched stuff just can't pack the way the linear stuff can, so it's highly amorphous.

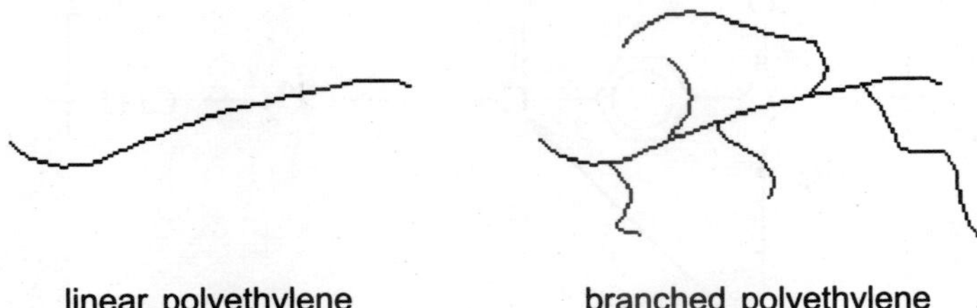

Crystallinity and Intermolecular Forces

Intermolecular forces can be a big help for a polymer if it wants to form crystals. A good example is nylon. You can see from the picture that the polar amide groups in the backbone chain of nylon(6,6) are strongly attracted to each other. They form strong hydrogen bonds. This strong binding holds crystals together.

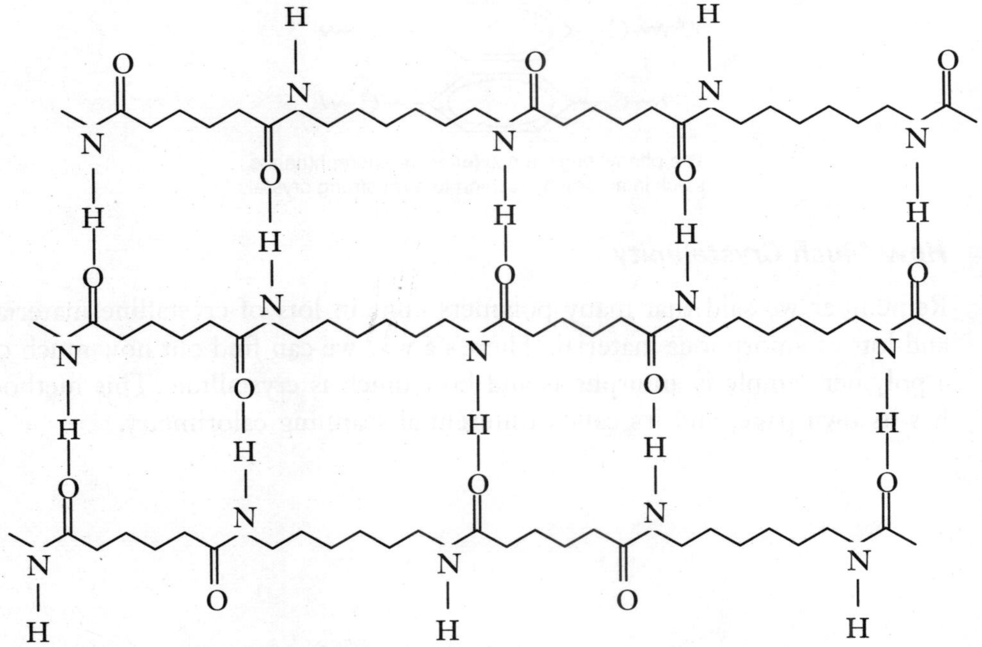

In nylon 6,6, the carbonyl oxygens and amide hydrogens can hydrogen bond with each other. This allows the chains to line up in an orderly fashion to form fibers.

Polyesters are another example. Let's look at the polyester we call poly(ethylene terephthalate).

$$\left[O - \overset{\overset{O}{\|}}{C} - C_6H_4 - \overset{\overset{O}{\|}}{C} - O - CH_2 - CH2 \right]_n$$

The polar ester groups in this poly(ethylene terephthalate) hold the polyester into strong crystals.

The polar ester groups make for strong crystals. In addition, the aromatic rings like to stack together in an orderly fashion, making the crystal even stronger.

The phenyl rings in poly(ethylene therephthalate stack in an orderly fashion to form strong crystals.

How Much Crystallinity

Remember we said that many polymers contain lots of crystalline material and lots of amorphous material. There's a way we can find out how much of a polymer sample is amorphous and how much is crystalline. This method has its own page, and it's called differential scanning calorimetry.

6

Polymer Material Science and Engineering

6.1 Polymer Materials and Properties

Introduction

A very wide range of polymer materials has been developed. This section briefly describes epoxies, the most versatile and pervasive of the thermosets, and some common thermoplastic and elastomeric materials.

Table 6.1 lists typical comparative mechanical properties for key polymers. It is important to realise that, within the range of grades that exist for a particular material, there can be significant differences in mechanical properties.

Table 6.1: Mechanical properties of typical polymeric materials.

Common name	*Chemical name*	*Specific gravity*	*Tensile strength (MPa)*	*Elongation at break (%)*	*Flexural modulus (GPa)*	*Max. operating temp. (°C)*
ABS (high impact)	acrylonitrile butadiene styrene	1.04	40	8	2.2	70
Acrylic	polymethyl methacrylate	1.18	70	2	2.9	55
Epoxies		1.2	70	3	3.0	130
LDPE	(low density) polyethylene	0.92	10	400	0.2	80
Polycarbonate		1.15	90	100	2.8	115
Polyester	e.g. polyethylene terephthalate	1.8	50	2	9.0	130
PTFE	polytetrafluoroethylene	2.1	25	200	0.5	180
PVC (flexible)	poly(vinyl chloride)	1.3	15	300	0.007	50
PVC (rigid)		1.4	50	60	3.0	90
Silicones		1.4	5	450		240

Source: Alvino 1994.

Note that the figures given are approximate mid-points, and for any material there will be a wide range of properties, depending on the formulation and polymerisation/cure conditions.

Polymer Materials

Epoxies

Epoxies first became commercially available in Germany in the 1930s, and by the 1950s began to be used as coatings and casting resins despite their high price. Today, PCBs using epoxy monopolise the marketplace, and epoxies are widely found in structural adhesives, structural composites and fibre-glass.

Epoxies may be cured at room temperature using amine catalysts, or cured at up to 180°C using amines or anhydrides to give superior thermal and physical properties. Most laminates are cured at around 150°C under pressure.

The most important epoxies are the epichlorohydrin-bisphenol-A types. Bisphenol-A is prepared by reacting phenol and acetone; epichlorohydrin is produced directly from the petroleum fraction propylene. In the presence of a catalyst, the two react to form a long chain molecule having carbon atoms and benzene rings interspersed along the backbone (Figure 6.1). In each molecular unit there are two pairs of carbon atoms, each pair of which shares an oxygen atom to form the epoxide group. It is this pair which opens in the presence of a curing agent to form the cross-linked, cured epoxy resin.

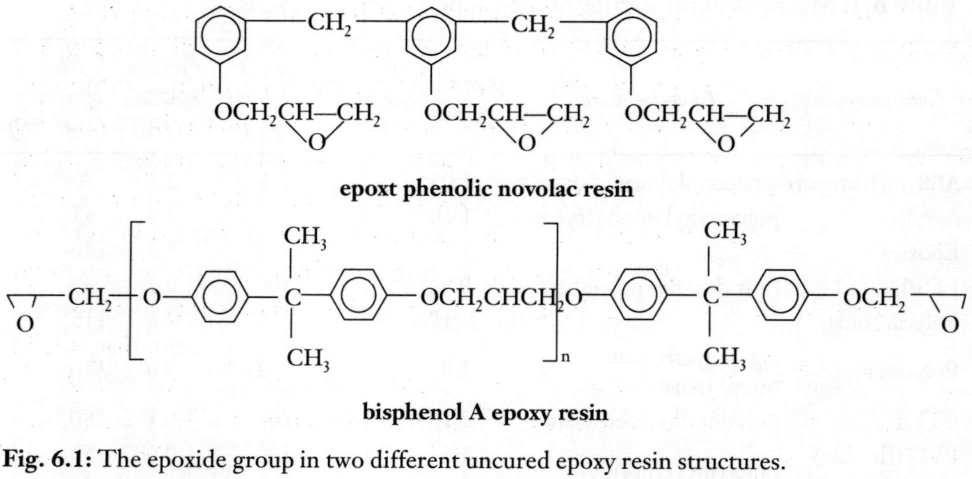

CH_2 CH_2

$OCH_2CH—CH_2$ O $OCH_2CH—CH_2$ O $OCH_2CH—CH_2$ O

epoxt phenolic novolac resin

CH_2 O CH_3 C CH_3 OCH_2CHCH_2O n CH_3 C CH_3 OCH_2 O

bisphenol A epoxy resin

Fig. 6.1: The epoxide group in two different uncured epoxy resin structures.

Acrylics

Acrylics are transparent, light-stable resins that have been manufactured since the early 1930s. The clarity and weather-resistance of cast acrylic explains its major commercial application for glazing, and acrylic fibres are used both for fabric and fibre optic applications.

Poly(methyl methacrylate) (PMMA) is a linear carbon-chain compound with a relatively low softening point, which has a higher strength than most thermoplastics, although it is brittle. It has low moisture absorption and is resistant to alkalis, dilute acids, detergents and greases, but will dissolve in alcohols, ketones and chlorinated solvents.

Silicones

Other polymers have a skeleton of carbon atoms: silicones are unique in that the skeleton is of silicon. Silicones range widely in molecular weight, and have a corresponding range of properties between light oils, greases and elastomers. Silicone resins cure to form solid but flexible materials which are impervious to moisture and have relatively high thermal conductivity.

Care has to be exercised in the selection and use of silicone elastomers:

- Certain types produce acetic acid during cure, which can have an adverse effect on some types of metallisation.
- Silicone fluids (monomers and low molecular weight polymers) are very effective release agents (which can make bonding problematic) and inhibit solder wetting.
- Acrylics and urethanes will dewet on silicone surfaces, so that silicone elastomers cannot be used in assemblies intended for subsequent conformal coating.

Additives

All thermoplastic, thermoset, and elastomeric materials are generally compounded before processing, with a variety of additives to modify the optical, mechanical or surface properties of the plastic. These include vulcanising and curing agents, accelerators, pigments, catalysts, antioxidants, plasticizers, and tack enhancers.

Some additives modify the polymer properties by physical means (plasticizers, lubricants, impact modifiers, fillers or pigments); others act by chemical change (heat stabilisers, antioxidants, UV stabilisers and flame retardants). All must be carefully chosen to be compatible with the processing method and the intended application of the finished product.

Some additives are organic and inorganic 'fillers'. These modify the mechanical properties of the structure, but are often primarily included to reduce cost. Adding filler to the resin is normally carried out at elevated temperatures, where surface tension and viscosity are both low. This encourages good wetting of the filler by the polymer, so that intimate contact between filler and polymer is maintained even when the material is deformed or exposed to water or other liquids or vapours.

In general, adding fine particles to polymers makes them more rigid, but also reduces their breaking strain and 'toughness', roughly in proportion to the volume of filler. Fibrous fillers behave differently: provided that the volume loading is sufficiently high, the length-diameter ratio sufficiently large, and the fibres themselves strong, the tensile strength, stiffness and flexural strength are determined largely by the filler rather than the resin. The properties may be markedly 'anisotropic', that is, the properties are not the same in all three axes. For example, glass mat reinforcement gives properties which are uniform in only two dimensions, not three.

A good reinforcing additive: is stiffer and stronger than the polymer matrix has good particle size, shape, and surface character for effective mechanical coupling to the matrix preserves the desirable qualities of the polymer matrix achieves the design objectives at lowest net cost.

Glass fibre is by far the most widely used fibre reinforcement, the result of having desirable properties for reinforcement and a moderate price (about the same as common plastics). The glasses used are all based on silica, with smaller quantities of other inorganic oxides. Compositions are identified by code letters: the cheapest, and most commonly used, is 'E-glass'; 'S-glass' has higher strength; 'C-glass' has improved resistance to attack by water and acids.

Self Assessment Questions

You are given a material which you are told is a 'B-stage' epoxy. Explain what this means. How would you expect its properties to differ from an epoxy moulding?

Mechanical properties of polymers

Density

Unreinforced plastics have densities in the range 830-2200 kg/m^3. This compares with values of 2700kg/m^3 for aluminium and 8000 kg/m^3 for stainless steel. Expanded polystyrene foam (used as a packaging and insulating material) can have a density as low as 10 kg/m^3, although structural foams usually have a density of at least 500 kg/m^3. The major advantage this low density gives plastics is that their strength/stiffness to weight ratios compare favourably with apparently much stronger materials. For example, on a stiffness to cost ratio there is little to chose between aluminium alloy, alloy steel, polypropylene and PVC. The choice of material can then be based on ease of manufacture and factors such as chemical or corrosion resistance and weatherability.

Strength and Stiffness

An important property of materials is their response to the application of a force, which can be classified into two main types of behaviour:

- ❖ Elastic materials return to their original shape once the force is removed.
- ❖ Plastic materials will not regain their shape and instead flow, behaving like a highly viscous liquid.
- ❖ Most materials demonstrate a combination of elastic and plastic behaviour, showing plastic behaviour after an 'elastic limit' has been exceeded. This behaviour is strongly dependent on temperature.

Plastics are generally stronger in compression than in other modes, and the values for tensile and flexural properties in Table 6.2 appear to restrict the use of unreinforced plastics in load bearing applications. However, their favourable strength to weight ratios and manufacturing benefits compared with metals make plastics preferable in many applications, giving the designer the challenge of compensating for low strength and stiffness by good engineering design. Usually this involves devising shapes which will give a sufficiently rigid construction, whilst using thin sections to facilitate easy moulding and fast production.

Table 6.2: Comparative mechanical properties of selected structural materials.

sss	*unreinforced plastics*	*aluminium*	*stainless steel*	*reinforced plastics*
tensile strength	10-100 MPa	170 MPa	740 MPa	100-1700 MPa
modulus	0.2-3.5 GPa	70 GPa	210 GPa	3-150 GPa

Source: Crawford 1985.

Of course the strength and stiffness of plastics can also be increased significantly by reinforcement (with glass or carbon fibres), and it is this approach which is adopted for the printed wiring board laminate.

Note that the yield strength of any plastic varies considerably with temperature: a strength range of greater than 2 : 1 over the temperature range -20°C to +70°C is typical. As the temperature is increased, the material becomes more flexible, and so for a given stress the material deforms more.

It should be kept in mind that short-term strength and stiffness values give a guide to the nature of the plastic, but should not be used in design calculations because plastics are sensitive to temperature and tend to creep and recover. As a general rule the long term strength of plastics decreases as the service temperature increases.

Toughness

Plastics exhibit a wide range of behaviour when subjected to impact forces. Whilst acrylic and polystyrene are brittle and fracture like glass, some plastics are very tough: polycarbonate is virtually unbreakable and is widely used in vandal-proof outdoor light fittings and safety helmets; nylon is used in industrial gears; ABS is used for crash padding in cars.

Unfortunately, comparing impact behaviour is difficult because different test methods use sample geometries with varying degrees of stress concentration. As the ranking of materials depends on the test method, designers have to be very careful when choosing a plastic on the basis of its tabulated impact strength, to ensure that the results quoted were obtained in a test which simulates closely the service conditions of the product.

In general, plastics will become brittle if they are subjected to sub-zero temperatures and/or contain severe stress concentrations, such as sharp corners and notches, and these conditions should be avoided whenever possible. Other factors known to lead to embrittlement include the use of excessively high melt pressures during injection moulding, which causes high residual stresses. Other areas of weakness in moulded plastics include weld lines and regions of highly oriented material, which will be strong in the orientation direction but relatively weak in the transverse direction. A common method of improving the impact strength of brittle plastics, such as polystyrene, is to add rubber fillers, but this is usually at the expense of strength and stiffness.

Fatigue

Two types of fatigue failure are observed:

- ❖ 'Creep failure' or 'static fatigue' which occurs when a thermoplastic is subjected to a constant load for a prolonged period. The higher the stress, the shorter will be the time to failure.
- ❖ More conventional fatigue failure, when parts are subjected to the repeated action of a relatively low stress which would not cause failure in a single application. Studies have shown that fatigue cracks can be initiated and propagated in plastics, just as in metals, although cracks generally develop within the wall thickness of the moulding rather than start from the surface.
- ❖ Depending on the application, failure may be the result of less than a complete fracture. For example, when stressed, transparent plastics develop crack-like features ('crazes') and other plastics may whiten, a change in appearance which may end their useful life.

Operating Temperature

The glass transition temperature (T_g) is the temperature at which the links in the polymer start to weaken and become more flexible. At temperatures above T_g, sections of the polymer backbone are relatively free to move, whereas below T_g their motion becomes frozen, with only small scale molecular motion remaining. Over a temperature range of only 5-10°C, the material thus becomes more rubber-like rather than rigid (or 'glass-like'). This phase transition is reversible and will leave no permanent effect on the polymer unless it is broken or allowed to distort whilst in the abnormal state.

A knowledge of T_g is essential in selecting polymers, which are compounded to produce the required T_g for the application. In general, elastomers have values of T_g well below room temperature, and structural polymers have values above room temperature. One example is of poly(vinyl chloride) (PVC), which is normally a brittle solid at room temperature, with a T_g of 83°C. It is therefore suitable for use for domestic cold water pipes, but not for hot water. However, by adding a small amount of plasticizer, the T_g can be lowered to –40°C, turning the PVC into a material which is soft and flexible at room temperature, and can be used for applications such as garden hoses. But even a plasticized PVC hose can become stiff and brittle in a severe winter, in the same way that natural rubber becomes brittle when cooled in liquid air, which is below its glass transition temperature of –70°C!

In general, the maximum continuous operating temperature is determined for thermoplastics by the heat distortion point, or for thermosets by the glass transition temperature (T_g), as above these temperatures the polymer is not sufficiently mechanically stable. The glass transition temperature is particularly significant in the case of:

- Thermoset materials used to make printed wiring boards, where the mechanical strength is lower above T_g, which can lead to distortion and warping during unsupported reflow soldering.
- Encapsulation resins, where exceeding T_g can cause excessive stresses on embedded components. This is caused by differences in the rate of expansion, as resins above T_g typically show large increases in TCE.

Elastomer Properties

Elastomers are affected by the environment more than other polymers. Thermal ageing usually increases cross-linking and the chain length of the elastomer, thus increasing stiffness and hardness and decreasing elongation. Radiation has a similar effect.

Elastomers are sensitive to oxidation and in particular to the effects of ozone. Ultraviolet radiation acts in a similar way to ionising radiation, so some elastomers do not weather well. Environmental effects are especially noted on highly stressed parts. Some elastomers are particularly affected by hydrolysis, and complete chemical reversion has been experienced, where the polymer depolymerises back to a liquid state.

Some changes occur also when elastomers are subjected to continued stress:

- ❖ Creep refers to a change in strain when stress is held constant.
- ❖ Compression set is creep that occurs when the elastomer has been held in compression.
- ❖ Permanent set is deformation remaining after a stress is released.

The hardness of elastomers is a measure of their resistance to deformation measured by pressing an instrument into the elastomer surface. Special instruments have been developed, the most common being the Shore durometer. Figure 6.2 shows the hardness of elastomers and plastics. On the Shore A scale, 0 is soft and 100 is hard. On this scale, a rubber band is 27, and a car tyre is 70, with squeegees being rather harder still (typically in the range 70 to 95).

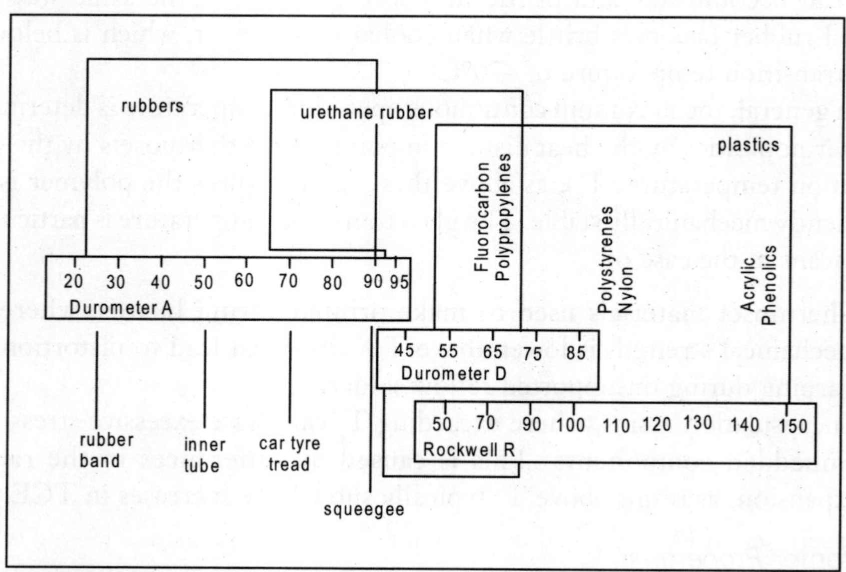

Source: Harper 1997.

Fig. 6.2: Hardness of rubber and plastics

Flammability

Flammability has little direct impact on any electrical or mechanical parameters (although additives can cause slight modification to the resin properties), but is extremely important from the point of view of safety. Polymers are used in many electronic devices, from radios to aircraft and the materials used in making the electronic circuits within these devices must not cause or dramatically assist combustion.

A cautionary Note on Nomenclature

Most people use the term 'inflammable' to refer to things which catch fire readily. The term 'flammable' is preferred in the technical literature, as being more correct, but means exactly the same.

The converse of both is 'non-flammable'. The most usual flammability tests are those specified by the American Underwriters' Laboratories (UL). For historical reasons, related to the use of PCBs in racked systems, these are carried out on test strips of standardised dimensions clamped vertically, with a defined flame applied for 30s and then removed. In 94V-0, the highest UL grade, the polymer has to self-extinguish within 10s, with no drips of flaming molten material. Materials with lower specifications (94V-1, 94V-2) are allowed longer to self-extinguish. The rationale behind the choice of time-scales is that heating caused by fault over-current is the most common source of ignition, and fuse protection will operate within 30s. Although it is not realistic to expect freedom from burning while a source of ignition is present, the polymer should not remain alight afterwards. The tests are applied both to materials such as base laminates and solder resists separately and to their combination in the completed PCB. The UL tests are practical tests on real products, and have been criticised on the grounds that they are biased towards testing PCBs rather than other components, results may be affected by factors such as surface roughness and sample thickness, and the test method is difficult to standardise and operator-dependent to some extent, being based on simple equipment. For this reason UL tests may not reflect the inherent flammability of the resins themselves. A supplementary method, preferred by bulk polymer suppliers, is the so-called Critical Oxygen Index (COI), also known as the Limiting Oxygen Index (LOI). This is the lowest percentage of oxygen in a nitrogen stream passing a specimen under test in which the specimen will remain burning under specified conditions. The technique requires extensive laboratory facilities, but is more reproducible than the UL test. However, the user will probably argue that the COI result requires interpretation, whereas UL test relate directly to the practical application.

For most applications, fire retardant additives are included in the resins used to make board laminates whence, for example, the name 'Fire Retardant version 4', or FR-4. Note that this term applies to the base material of which the PCB is made, and does not describe the additives.

However, such additives (most commonly brominated compounds) frequently greatly increase the harmful by-products of combustion. Since the Falklands War and Kings Cross fire, end-users have become aware of the dangers from such smoke and fumes. There is consequently a fine balance to maintain between meeting a tight flammability specification and avoiding excessive use of possibly harmful materials.

6.2 Polymer Composites as Construction Material

Project Objective

To collate, interpret, analyse and disseminate recent international research knowledge and development of polymer composite materials in construction to encourage uptake.

Polymer Composites in Construction

About 30 per cent of all polymers produced each year are used in the civil engineering and building industries. Polymers offer many advantages over conventional materials including lightness, resilience to corrosion and ease of processing. They can be combined with fibres to form composites which have enhanced properties, enabling them to be used as structural members and units. Polymer composites can be used in many different forms ranging from structural composites in the construction industry to the high technology composites of the aerospace and space satellite industries.

Polymer composites were first developed during the 1940's, for military and aerospace applications. Considerable advances have been made since then in the use of this material and applications developed in the construction sector. Load bearing and infill panels have been manufactured using composites. Complete structures have been fabricated where units manufactured from glass-reinforced polyester are connected together to form the complete system in which the shape provides the rigidity. Glass-reinforced plastics have been used in many other applications including pressure pipes, tank liners, and roofs.

In contrast to the design procedures used for the more traditional construction materials, those for polymer composite materials in structural applications require greater development effort and a wider understanding of the material. The material properties of the final component are the result of a design process that considers many factors which are characterised by the anisotropic behaviour of the material and cover the micro-mechanical, elasticity, strength and stability properties. These properties are influenced by manufacturing techniques, environmental exposure and loading histories. Designing with composites is thus an interactive process between the design of the constituent materials used, the design of the composite material and an understanding of the manufacturing technique for the composite component.

In the last decade, polymer composites have found application in the construction sector in areas such as bridge repair, bridge design, mooring cables, structural strengthening and stand-alone components.

This project, funded by DTI, was launched in July 2001 with the following key objectives:

- ❖ Collate, interpret and analyse international research knowledge of FRP materials in construction from 1990 onwards.
- ❖ Repackage the key information into informative and digestible forms, to establish and highlight synergy's with traditional construction materials for achieving design and cost efficiencies.

The principal target audience for this project is individuals involved in application of building materials within the construction industry that have little or no knowledge of composites or are for whatever reason predisposed against them. Dissemination plays a major role within the project and the team will be working closely with NGCC to ensure this is carried out efficiently and effectively. Trend2000 and BRE will lead the project with support from Brookes Stacey Randall, Concrete society, Fibreforce Composites ltd, Tony Gee and Partners, University of Sheffield, and Saint-Gobain Vetrotex. The project will run for 18 months from its start date.

6.3 Department of Polymer Materials

Research

The Department pursues basic research in polymer materials, their preparation and properties. Main attention is paid to: (i) study of polymer miscibility and interfacial tension in polymer systems, (ii) microrheological description and experimental investigation of the phase structure formation and evolution in immiscible polymer blends, (iii) study of the effect of reactive and copolymer-based compatibilization on structure and properties of polymer blends, (iv) development of rules for prediction of properties of polymer blends from the knowledge of their phase structure and properties of components, (v) study of the course of polymer crystallization, and (vi) investigation of degradation and stabilization of polymer materials. The results are utilized in studies of recycling commingled plastic waste and in development of new materials based on polymer blends.

During the year 2000, the research was focused mainly on the following problems:

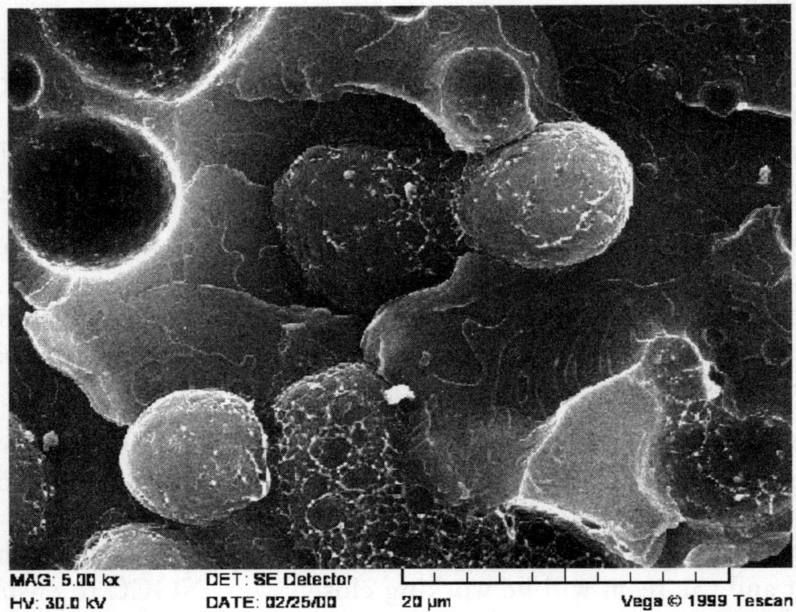

- ❖ Elucidation of differences between compatibilization efficiency of styrene-butadiene block copolymers in polystyrene/polybutadiene, polystyrene/polyethylene and polystyrene/polypropylene blends.
- ❖ Study of creep in polypropylene/elastomer/poly(styrene-*co*-acrylonitrile) blends and its description by the model based on combination of the equivalent blocks model and modified equations of the percolation theory.
- ❖ Effect of composition of the combined compatibilizer, developed in the department, on its compatibilization efficiency in real municipal plastics waste.
- ❖ Analysis of the effect of breakup frequency of droplets in polymer blends on their average size in steady flow.
- ❖ Effect of the g-ray dose on selected properties of ultra-high-molecular-weight polyethylene. Theory of the open association suitable for description of reversible physical gels. Description of the shear-flow-induced coalescence in immiscible polymer blends.

Organisation Structure

- ❖ Working group of Polymer Materials Development and Recycling.
- ❖ Working group of Thermodynamics and Rheology.

- Working group of Electronic Phenomena.
- Joint laboratory of polymer materials of the Institute of Macromolecular Chemistry, Academy of Sciences of the Czech Republic and Tomáš Bat'a University Zlín.

6.4 Advanced Polymer Materials Inc.

Advanced Polymer Materials Inc. is a Montreal based manufacturing company with R&D capability. We produce innovative and unique biodegradable and biocompatible polymer materials to fill the need for scientific advancement in pharmacy and medicine. We also provide custom synthesized polymers upon demand and according to our customers' specifications.

We are committed to exceeding our customers' expectations, whether it is in providing the highest quality and reliable polymer materials, or in our unique services such as consultancy and developing new technologies.

Business

We supply the following materials, technologies and services.

Materials & technology	*Application*
PHAs and low molecular weight PHAs, e.g. Low molecular weight PHBs	Biodegradable materials, functional blocks for graft polymers, tissue engineering. Microspheres, nano particles for controlled release formulation
Functionalized PEGs e.g. mPEG-COOH, mPEG-NH_2, etc.	For PEGylation and conjugation
Biodegradable copolymer materials, e.g., PEG/PHB, PEG/PLA, PEG/PCL, etc. block copolymers	Drug delivery, tissue engineering, food additives, cosmetics, etc. Biodegradable blending materials Nano medicine
Block copolymers of ethylene oxide and propylene oxide, e.g., PEO-PPO, PPO-PEO-PPO, etc. block copolymers	Supplementary to the Pluronic copolymers. Micelle and Gelation agents
Block copolymers of ethylene oxide and 1,2-butylene oxide, e.g., PEO-PBO, PBO-PEO-PBO block copolymers Block copolymers of ethylene oxide and styrene oxide, e.g. PEO-PSO, PSO-PEO-PSO block copolymers	Better solubilization and more stable micelle systems Templates for preparation of nano-materials. Templates for preparation of mesoporous materials.
Cyclic polymers and copolymers, e.g., cyclo-PEG, cyclo-PEO/PPO copolymers, etc.	R&D research in polymer morphologies and polymer behaviors.
Contract research	Special applications, new polymer systems and technologies.
Instrument service	Polymer analysis and characterization

6.5 Product Catalogue

1. Biodegradable Polymers and Copolymers
   (*a*) Poly((R)-3-Hydroxyalkanoate)s (PHA)s, Natural Polymers.
   (*b*) Low Molecular Weight PHAs (LMW PHA).

2. Biodegradable Block Copolymers
   (*a*) PHB-B-PEG Block Copolymers (3-201).
   (*b*) PHB/V-B-PEG Block Copolymers (3-202)
   (*c*) PHB-B-PEG-B-PHB BAB Type Triblock Copolymers (4-201)
   (*d*) PHB/V-B-PEG-B-PHB/V BAB Type Triblock Copolymers (4-202)
   (*e*) Block Copolymers of Poly(ethylene glycol) and Poly(caprolactone) (PEG-b-PCL or PCL-b-PEG) (5-01)
   (*f*) PCL-b-PEG-b-PCL, BAB type triblock copolymers) (5-02)
   (*g*) Block Copolymers of Poly(ethylene glycol) and Poly(Lactide) (PEG-b-PLA or PLA-b-PEG) (6-01)
   (*h*) PLA-b-PEG-b-PLA, BAB type triblock copolymers (6-02)

3. Oxirane Block Copolymers:
   (*a*) PEO/PPO AB Type Diblock Copolymers (PEO-b-PPO or PPO-b-PEO) (7-01)
   (*b*) PEO/PPO BAB Type Triblock Copolymer (7-03)
   (*c*) PEO/PBO AB Type Diblock Copolymers (PEO-b-PBO or PBO-b-PEO) (8-01)
   (*d*) PEO/PBO BAB Type Triblock Copolymers (8-03)
   (*e*) PEO/PSO AB Type Diblock Copolymers (PEO-b-PSO or PSO-b-PEO) (9-01)
   (*f*) PEO/PSO BAB Type Triblock Copolymers (9-03)

4. Cyclic Polymers and Copolymers:
   (*a*) Cyclic Poly(Oxyethylene) With Acetal Linkage (10-01)
   (*b*) Cyclic Poly(Oxyethylene) Large Crown Ether (10-02)
   (*c*) Cyclic Block Copolymers Of Ethylene Oxide and 1,2-Butylene Oxide (10-03)
   (*d*) Cyclic Block Copolymers Of Ethylene Oxide and Propylene Oxide (10-04)

6.6 Working Group of Polymer Materials Development and Recycling

Research

The activity of the Group is focused on several topics having character of basic research as well as on more practical problems connected with development and/or modification of multiphase polymer materials and with recycling of plastics waste.

Development and/or Modification of Multiphase Polymer Materials

The applied research projects deal with blends of polystyrene/polyolefin, polyamide/ABS, polyethylene/polypropylene, polycarbonate/SAN and others using non-reactive as well as reactive compatibilizers. The aim of these projects is attaining balanced combination of end-use properties such as barrier mechanical characteristics, chemical resistance—toughness, etc.

Special attention is devoted to recycling of commingled plastics waste using special reactive compatibilizing systems leading to high-impact materials, preserving thermoplastic behaviour.

Current research topics also encompass:

- Long-term degradation and aging experiments with polyolefins, styrene polymers and their blends, estimation of residual oxidation stability of aged samples and lifetime predictions, as well as efficiency evaluation of various stabilizer systems.
- Correlation of the extent of oxidative deterioration with changes in processability in connection with successful recycling of polymers; contribution of restabilization.
- Exploitation of thermoanalytical methods (DSC, TGA) for characterization of homopolymers, polymer blends, composites and block copolymers.

6.7 Imaging and Mapping Trends For Polymer Materials with Veeco Atomic Force Microscopes

Background

Initially introduced as an accessory to , atomic force microscopy (AFM) has become an advanced and most valuable scanning probe technique broadly used in academic and industrial research. A superior feature of AFM—the high-resolution visualization of surfaces—defines most of its applications. On one hand, AFM microscopes are employed for imaging of individual molecules and manipulations of nanometer-scale objects. On the other hand, they are widely applied for quality control in a number of industries where is used for measurements of surface roughness and for quantitative examination of shapes/profiles of technologically important surface structures, such as DVD and CD patterns.

Over the past decade, many researchers working with soft matter (biological objects and polymers) have recognized the importance of this method. Nanometer-scale features as small as individual macromolecules, their assemblies, and larger-scale morphologies, can be easily recognized in AFM images. In addition, the fact that the stiffness of AFM probes is comparable to the stiffness of polymer materials makes it possible to distinguish sample locations with different mechanical properties. In this way, various components of heterogeneous polymer materials can be identified in the images.

For many years, examination of polymer samples with AFM was performed at ambient conditions and under liquids. These studies have been recently extended to different temperatures. In this manner, AFM can be used for in-situ monitoring of structural changes induced by thermal transitions, and for visualization of structure transformations caused by swelling and other effects. This application note presents several novel applications to polymers.

Optimization of AFM Experiments

Compared to other microscopic techniques, is still in its infancy. The capabilities of AFM, especially in studies of soft materials, are continuously expanding with the ongoing development of instrumentation and the increasing practical experience of researchers. Recent instrumental developments include a design of more precise actuators and more sensitive detectors, new operation modes, and specialized probes. There is also room for further improvement of overall microscope design and, particularly, lowering its thermal drift.

In practical AFM applications, it is well established that the tapping mode imaging technique is preferable to the contact mode technique. Depending on imaging needs, the experimental conditions should be adjusted either for high resolution and gentle profiling of surface features or for high-contrast visualization of different components of heterogeneous polymer systems (polymer blends, block copolymers, polymer composites). In tapping mode, the level of the tip-sample force interactions is varied by changing the amplitude of free-oscillating probe (A0) and set-point amplitude (Asp), as well as by using probes with different stiffness. For example, low-force imaging (light tapping) is achieved by applying soft probes (stiffness as low as 0.3 N/m), using small A_0 and A_{sp} close to A_0. These conditions are favorable for non-destructive, high-resolution imaging of soft materials.

A good criterion of light tapping is imaging, at which the phase of the interacting probe is close to its phase in free oscillation. An operation at elevated tip-sample forces, which can be achieved by lowering A_{sp}, increasing A_0, and by using stiff probes, is optimal for compositional imaging of individual components in heterogeneous materials. In many cases, individual components are distinguished by different contrast in phase images. For compositional imaging of multicomponent materials with rubbery components, the most pronounced phase contrast is usually achieved at hard tapping conditions (A_0 = 80-100nm, A_{sp} = 40-50nm, probes with ~40 N/m stiffness). In hard tapping of polymer samples with rubbery top layers, the tip depresses the samples and even penetrates into them thus it probes sub-surface structures. The tip penetration can reach hundreds of nanometers when imaging of rubber samples is performed with A_0 higher than 100nm and A_{sp} as small as 5nm.

High-Resolution Imaging of Polymer Structures

At present, the question of ultimate resolution is still an open one. Since the beginning, AFM images obtained in contact mode have revealed crystallographic lattices of surfaces of organic and inorganic compounds but were unable to show isolated molecular scale defects (true molecular resolution). The use of oscillatory AFM mode in UHV allowed imaging with true atomic resolution on a number of crystalline surfaces, and even sub-angstrom imaging was achieved. Until recently, true molecular resolution in tapping mode in air was problematic. The use of novel ultra-sharp probes and an AFM instrument with low thermal drift makes this possible.

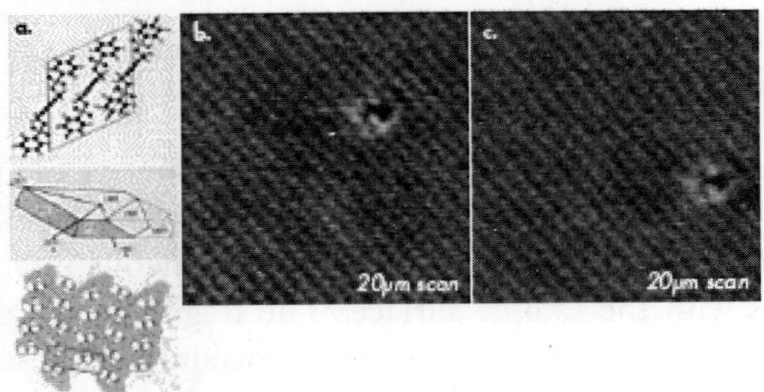

Fig. 6.3: (a) A sketch of chemical structure, a crystal shape and crystallographic order at the bc plane for polydiacetylene (PDA) used for high-resolution AFM imaging. (b)-(c) Height images of the bc surface of PDA crystal obtained in tapping mode in air. Carbon spike probe is used in the experiments. The images show an ordered packing of surface groups and one molecular-scale defect. The images were recorded at the same location within a 10 min. interval. The shift of the defect is related to thermal drift of ~0.6nm/min.

The example of tapping mode imaging with true molecular resolution, which was obtained in studies of polydiacetylene (PDA) crystal, is demonstrated in Figure 6.3. This polymer has a chemical structure presented by a sketch. The top surface (the bc plane) of its crystal is formed by terminal molecular segments of the chain side groups, which are packed in a rectangular cell with the main parameters: b = 0.491nm and c = 1.410nm. This image of PDA crystal shows not only the regular lattice, which is consistent with the crystallographic structure of the bc surface, but also a molecular-size defect. The image was obtained with an ultra-sharp probe, which is prepared by a plasma-assisted deposition of carbon spikes on the apex of a Si etched probe. The value of this probe is manifested not only on surfaces with an expected molecular- or atomic scale order but also in high-resolution imaging of less-ordered surfaces with corrugations and pores in the nanometer scale. The TEM micrographs of this probe, a routinely used Si etched probe, and a novel diamond probe are collected in Figures 2a-c. Such probes with sub-10nm apex size are invaluable for molecular-scale imaging.

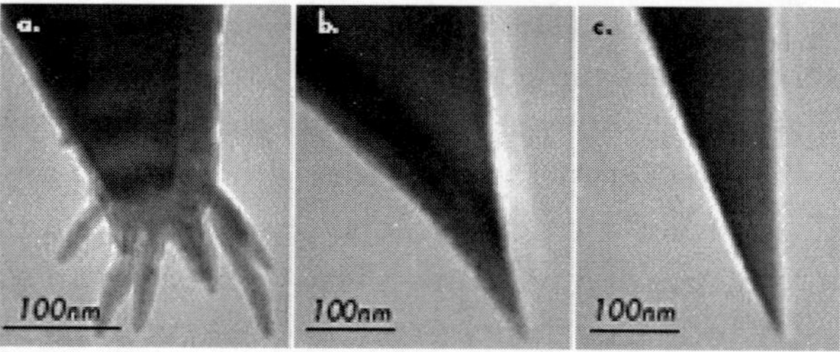

Fig. 6.4: TEM micrographs of AFM probes. (a) Carbon spikes grown at the apex of Si etched probe; (b) Etched Si probe; (c) Diamond tip. TEM micrographs and diamond tip—courtesy of B. Mesa (MicroStar Technology).

A presence of a number of spikes at the extremity of the probes (Figure 6.4a) imposes definite restrictions on their use. Relatively flat samples should be studied with such probes to avoid the simultaneous interactions of the multiple spikes with the sample surface. The fragile nature of sharp spikes also requires gentle engage and imaging procedures. These prerequisites are less demanding in the case of Si etched probes and diamond probes, whose apex dimensions can be in the sub-10nm range.

Besides the use of sharp probes, an AFM microscope with low thermal drift (below 0.5nm/min) is important for molecular-scale imaging in tapping mode because of the slow scanning inherent to this method. With this lowthermal-drift instrument, molecular-scale resolution was also achieved in images of an oriented polytetrafluoroethylene (PTFE) layer and liquid crystalline material using a regular Si etched probe (Figure 6.4b) and a diamond probe (Figure 6.4c). The high-resolution images of PTFE are shown in Figures 6.5a-b. Flat crystalline terraces with steps of 0.56nm in height are seen at the surface of this layer (Figure 6.5a), which was prepared by hot rubbing on a glass substrate.

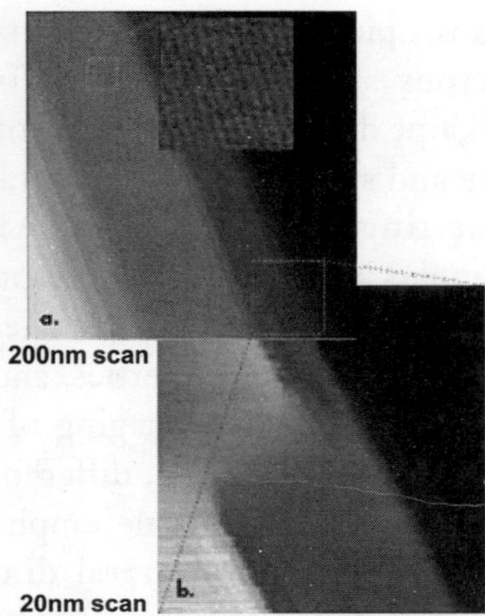

Fig. 6.5: (a) Height image of PTFE layer obtained in tapping mode in air. Higher magnification images shown in the top part of Figures 3a-b were obtained at the locations marked with dotted lines. Etched Si probe was used in the experiments.

This indicates that the surface is formed of single molecular sheets. A high resolution image of the 10-nm location (top, left) shows individual polymer chains with 0.56-nm spacing. There are also a number of defects, some of which are magnified in Figure 6.5b. The dislocation defects marked by the circles are made of PTFE chains, which extend from one molecular sheet to another. One more example of molecular-scale imaging is presented in Figure 6.6, in which images of an annealed layer of star-shaped molecules on Si substrate are collected. The morphology pattern, which is seen in the large-scale image in Figure 6.6a, is common for liquid crystalline materials. On flat domains, one can distinguish a crystalline order with the almost orthogonal lattice, which is characterized by the repeat distances of 3.4nm and 3.6nm. A comparison of the AFM images with the results of X-ray analysis will help reveal the molecular packing in this layer.

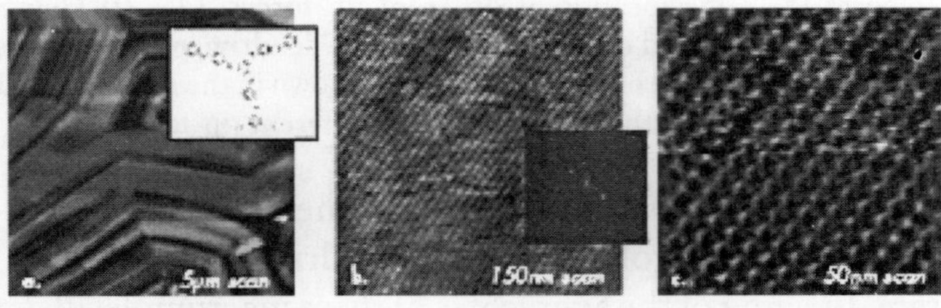

Fig. 6.6: (a)-(c) Height images obtained on surface of liquid crystalline material in tapping mode in air. A sketch of chemical structure of disk-shaped star molecules forming this material is shown in the insert in (a). 2D FFT power spectrum of the image in (b) is shown in the insert. A lower half of the image in (c) is shown after filtering. Etched Si probe was used in the experiments.

Microscopic and nanoscopic objects are recognized in images by more than their specific shapes and dimensions. Visualization of single macromolecules, which adopt different molecular conformations depending on a deposition procedure and specifics of their interactions with a substrate, is one of the most exciting AFM features. AFM studies of DNA macromolecules, for example, became the routine procedure more than 10 years ago. Imaging at different tip-forces often leads to image-contrast variations that reflect local mechanical properties, and this procedure can be applied to single macromolecules. When imaging of DNA macromolecules (Figures 5b-c) placed on mica is performed at different forces, the contrast of the structure representing the macromolecule emphasizes either its overall shape or a more rigid core, which has lateral dimensions more closely corresponding to the expected size of DNA.

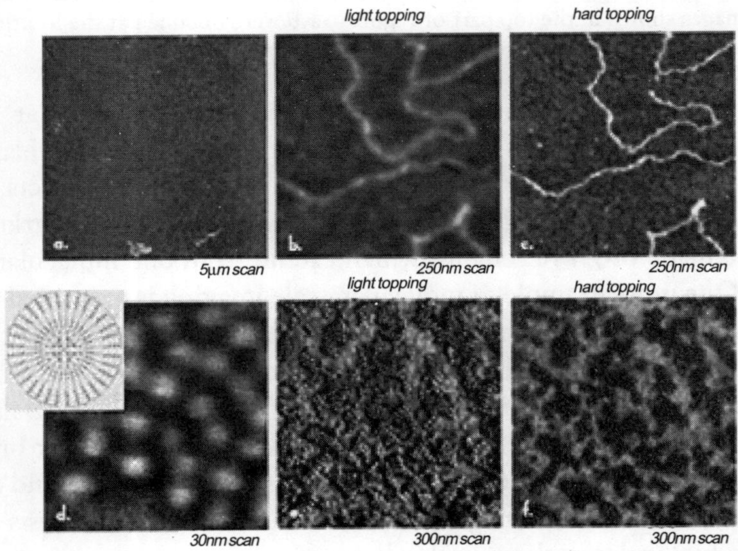

Fig. 6.7: (a)-(c) Height images of DNA macromolecules on mica. Images in (b) and (c) were obtained at the same location at different tip-forces. (d)—(f) Height images of macromolecules of liquid crystalline carbosilane dendrimer on Si substrate. A sketch of the chemical structure of these molecules is shown in the insert in (d). Images in (e) and (f) were obtained at the same location at different tip-forces.

In another example (Figures 6.7d-f), spheres 5nm in size represent individual macromolecules of carbosilane dendrimer, which were deposited on Si substrate. Imaging of the aggregates of these macromolecules at different forces shows that the exterior of the macromolecules changes its contrast from bright to dark as the tip-force increases, Figures 6.7e-f. At the same conditions, the core of the macromolecules remains bright, which indicates its higher stiffness compared to the periphery of the molecules. This result corresponds to the higher density of the central portion of the molecules, which was confirmed by molecular dynamics simulations. In both cases, the image changes induced by tip-force are reversible and reflect variations of local mechanical properties. A use of AFM for studies of individual synthetic polymer molecules was first demonstrated by visualization of single chains of poly(styrene)-b-poly(methylmethacrylate) block copolymer and by monitoring changes of their conformation caused by humidity variations. This work was followed by a rapidly increasing number of applications in which single dendrimer molecules, macromolecular brushes, and polymers with mini-dendritic groups were observed with . For such studies, single macromolecules are typically deposited on selected substrates (mica, silicon wafer, glass, graphite, etc.) from very dilute solutions.

The choice of substrate is essential for fixing macromolecules and unravelling polymer chains. The unraveling might be assisted by epitaxy of terminal alkyl groups to graphite, as it has been shown in studies of alkanes and polymer molecules with mini-dendritic side groups. Imaging of single macromolecules is not a trivial procedure and one can expect difficulties in deducing the correct size of individual macromolecules from AFM images. While imaging isolated nanometer-scale objects on a substrate, it is rather hard to avoid a convolution of the tip shape with the shape of a macromolecule, as well as a possible tip-induced deformation of single macromolecules. These effects should be taken into account during analysis of the width and height of the isolated extended chain from the images, yet the overall shape appearance and contour length of these macromolecules are reproduced in the images more correctly. Therefore, the images can be used for evaluation of the macromolecule conformation on different substrates and for construction of molecular length histograms, which provide direct information about molecular weight distribution.

An example of how AFM can be applied for conformation analysis of macromolecules on different substrates and in bulk was obtained in studies of macromolecules of polyphenylacetylene with mini-dendritic groups (Figure 6.8a).

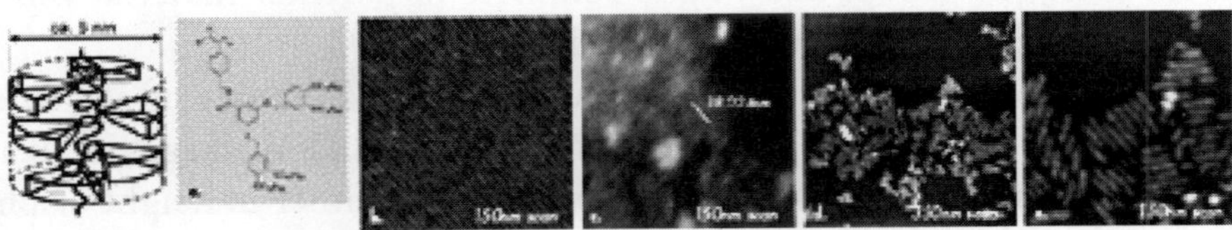

Fig. 6.8: (a) Chemical structure of polyphenylacetylene with mini-dendritic groups and a sketch of the appearance of these molecules in bulk based on the X-ray analysis. Height images of these macromolecules on the fracture surface of the bulk polymer sample (b), in a single molecules domain on mica (c) and on graphite (d) and (e). The number in the insert of (c) indicates a length of four molecules marked with a white line in the image. The red arrows in the image in (d) show one of the molecular stacks on graphite.

In bulk, these macromolecules are arranged in a hexagonal lattice and the diameter of individual chains, which are formed by the core wrapped in a "jacket" of mini-dendritic groups, is around 5 nm (X-ray data). The ordered chain layers, which were observed with AFM on the fracture surface of a bulk sample, exhibit a step height (4.5nm) equal to the macromolecule's diameter. A spacing of 4.9nm was found between the single chains seen in the image of a top surface of this sample (Figure 6.8b). These dimensions are close to the chain diameter (5.0 nm) determined with the X-ray analysis.

The AFM images of the polymer chains on mica and graphite are shown in Figures 6c-d. The chain spacing of ~4.5nm, which was determined from the image of the polymer aggregate on mica, is similar with that found in the molecular layers on the fracture surface. When the macromolecules were deposited on graphite, their epitaxy led to a chain alignment along the main crystallographic axes of the substrate, Figure 6.8d. The width of the macromolecules on graphite as measured in three different stacks (one of which is indicated by red arrows) in this image is larger (6.5–6.6nm) than on mica. The height of the macromolecules on graphite (~0.79nm) is also substantially reduced compared to that on mica and on the fracture surface. These observations suggest that the polymer chains are flattened on graphite but retain an almost cylindrical shape on mica.

Self-assembly of macromolecules on a substrate takes place as the concentration of their solution, which is used for the deposition, increases. Domains of packed macromolecules on substrates can be more easily observed in AFM because they can withstand the tip-force better than single macromolecules. The main questions that are raised from the analysis of AFM images of macromolecular self assemblies on different substrates concern the interplay of macromolecular interactions within the polymer layer and those between macromolecules and substrates. The latter typically influence macromolecular order in one or several layers lying immediately on the substrate. Intermolecular interactions within the layers can originate from different mesogen groups and might dominate in polymer material with no relevance to the layer proximity to the substrate.

Several interesting results were obtained in AFM studies of alkane and polyethylene (PE) layers on graphite. What is seen in AFM images of alkanes is an alternation of the strips with different contrast either in height or phase images. The images of $C_{18}H_{38}$, $C_{36}H_{74}$, and $C_{60}H_{122}$ alkanes in Figures 6.9a-c demonstrate that the width of the strips coincides with the length of the polymer chains. So far, AFM images showing individual alkane chains are not known, whereas visualization of individual $–CH_{2-}$ groups is usually achieved in STM images of alkane layers at the liquid-solid interface with graphite. In AFM images, arrays of the end $-CH_3$ groups at the lamellar edges are typically seen as darker lines. This assignment is confirmed by imaging of $C_{390}H_{782}$ layers (Figures 6.10a-d) where the reversible variations in the height and phase images were observed as the tip-sample force interactions were varied.

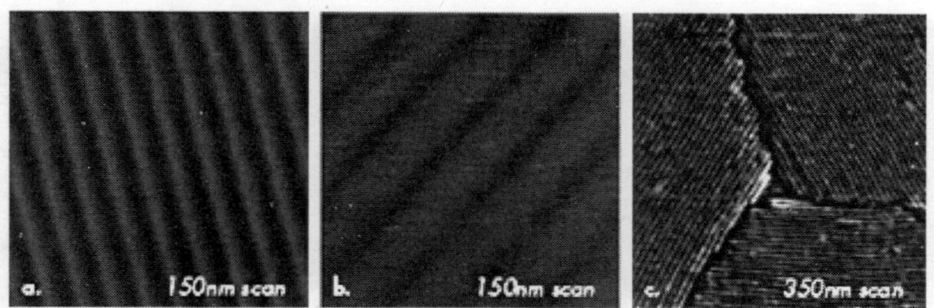

Fig. 6.9: (a)-(c) Phase images of $C_{18}H_{38}$, $C_{36}H_{74}$ and $C_{60}H_{122}$ alkane layers on graphite.

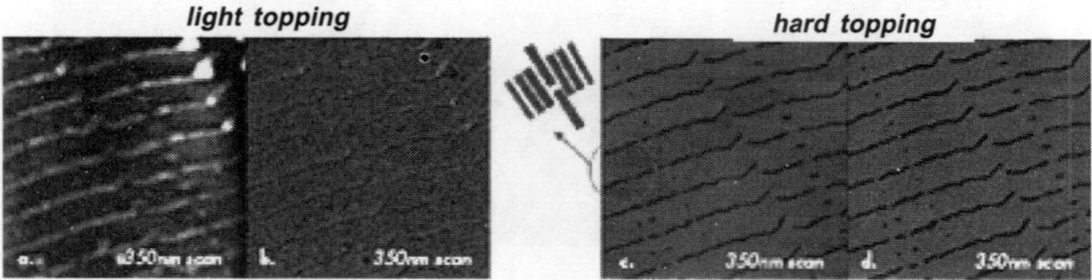

Fig. 6.10: (a)-(b) Height (left) and phase (right) images of ultra long alkane $C_{390}H_{782}$ on graphite obtained in light tapping. (c)-(d) Height (left) and phase (right) images of the same location as in (a) and (b) obtained in hard tapping. All images were obtained at T=130°C. One of the local defects related to the alkane molecules forming a bridge between neighbouring lamellar ribbons is indicated with a circle in (c).

In light tapping, the height image represents the surface topography; therefore, the sites of $-CH_3$ end groups of the alkane, which are bulkier than $-CH_{2-}$ groups, are seen elevated. A negligible contrast of the corresponding phase image confirms low tip-sample force interactions. When the tip-force increases (hard tapping), a possible tip deformation of regions of more mobile $-CH_3$ groups makes these locations appear darker in height and phase images; thus, the images represent lamellar order of the alkane layers on graphite and help visualize various defects. One such defect, which is related to alkane molecules forming a bridge between neighbouring lamellar ribbons, is pointed out in Figure 6.10d.

Studies of polymer samples at elevated temperatures can yield invaluable information concerning the organization of adsorbate layers on different substrates. Particularly, our work on high-temperature imaging of layers of alkanes of different lengths (from $C_{18}H_{38}$ to $C_{390}H_{782}$) and polyethylene (PE) on graphite has shown that the order at the melt-substrate interface is retained at temperatures substantially exceeding the melting temperatures of the related crystals. In such measurements, the tip penetrates through a melt and reaches the lamellar layer lying immediately on the substrate. The structural rearrangement proceeds at high temperatures in the fashion shown by images of $C_{390}H_{782}$ alkane and PE in Figures 9a–f.

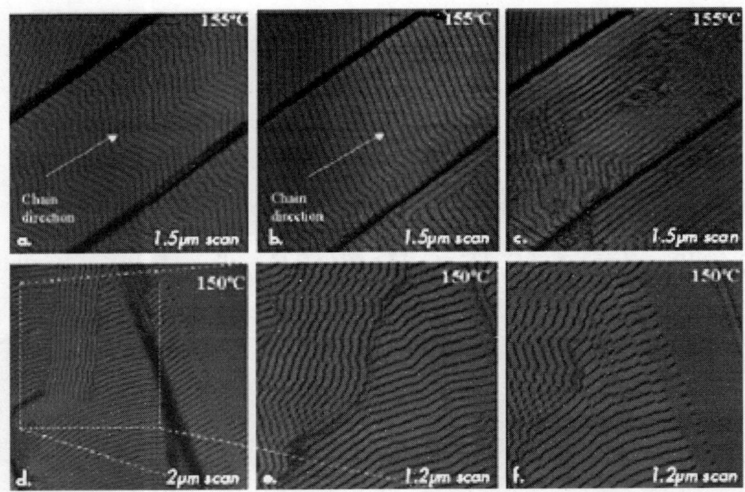

Fig. 6.11: (a)-(c) Height images of ultra long $C_{390}H_{782}$ layers on graphite at 155°C. (d)-(f) Height images of polyethylene layers on graphite at 150°C.

In the case of the $C_{390}H_{782}$ layer (Figures 9a–c), after the temperature reached 155°C, zigzag-shaped lamellae of the middle terrace first changed to a straighter shape and then reoriented along the terrace edges. These alternations reflect the interplay of several interactions between the molecules within the layer and between the alkane molecules and the substrate. More bulky –CH_3 end groups enforce the sliding of chains with respect to each other that caused the zigzag shape of the lamellae. With the temperature increase, molecular mobility was enhanced and enough space become available for the end groups of the neighbouring chains. This might explain the straightening of the lamellae. The increased molecular mobility also provoked a loss of registry of the alkane chains with one of the main directions of the graphite. As a result, the lamellar ribbons align preferentially along the terrace edges.

In the case of the PE layer (Figures 9d–f), the lamellar domains, which exhibited a well-defined orientation along the principal axes of the substrate (left image), gradually lost this epitaxial arrangement due to further increased molecular mobility (middle and right images). Ongoing isotropization of material, a loss of the lamellar order (see the right top corner of the right image), and breakage of individual lamellae into blocks is the result of local chain motion followed in this process.

Mapping of Multicomponent Polymer Materials

Multicomponent polymer materials are widely used in many industries because by appropriate mixing of different polymers and fillers one can design materials with desirable properties. The structure-property relationship in such materials is difficult to understand without microscopic analysis. AFM is very helpful in this analysis at scales from hundreds of microns to nanometers. The specific shape of individual components as well as variations in their mechanical and electric properties allow them to be distinguished from one another in AFMM images.

The images of two nanocomposites, polyparaxylylene with Si particles and polypropylene with clay, are shown in Figures 10a–b. Si particles, which are seen as small spheres with radii varying in the 5–10nm range, are dispersed in the polymer matrix. Small sheets of layered clay mineral, some of them in aggregates, are seen as edge-on structures in several locations in Figure 6.12b. The presence of aggregates indicates that the exfoliation process, which could separate the filler into multiple sheets with thickness down to molecular size, was not optimized.

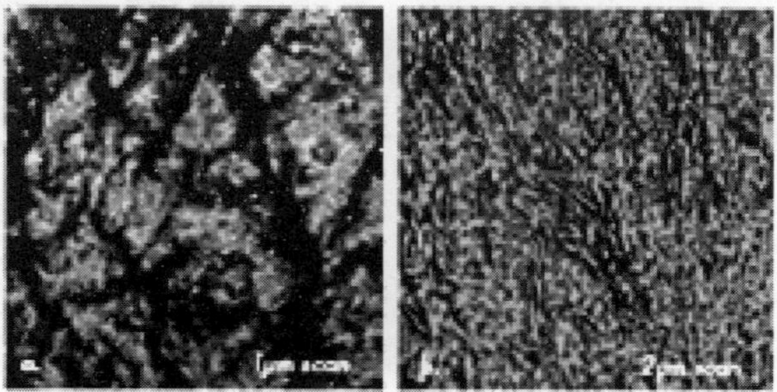

Fig. 6.12: Phase images of polyparaxylylene filled with Ag particles (a) and polypropylene filled with clay particles (b).

Further efforts towards high-resolution observations of the individual nanoparticles in the sub-100-nm images will be quite desirable not only for visualizing their distribution in the matrix but also for determining specific shapes of nanoparticles in multicomponent materials and their proper assignment, which is rather difficult now.

Visualization of the microphase separated morphology (spherical, cylindrical, lamellar, or micellar) of block copolymers, which is characterized by structural parameters in the 5-50 nm range, is one of the routine procedures in AFM applications. Domains of different components, which are formed as the result of phase separation of polymer blends, are typically of a larger size. AFM provides real-space observation of different coarsening processes and allows direct quantification of phase volume fractions and computation of domain interface curvatures of various morphologies. The contrast variations in images of block copolymers and polymer blends are related to differences in the mechanical properties of the component. AFM allows studies of local mechanical properties at scales down to sub-micron by examining force curves in different sample locations.

The images in Figure 6.13 show that macroscopic testing of polymer properties could be extended to the nanoscale, which is important for developments in nanotechnology. The force curves present deflection-versus distance (DvZ) or amplitude versus distance (AvZ) dependencies. In applications for soft materials, DvZ curves are often recorded while the tip penetrates a sample and actually damages the surface layer.

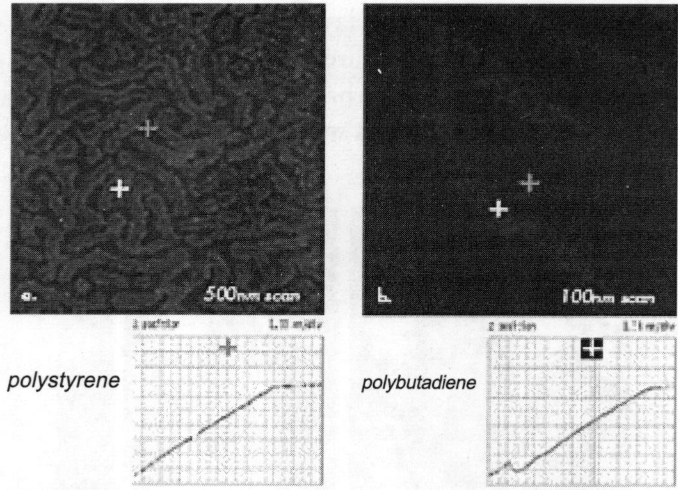

Fig. 6.13: (a)-(b) Force Volume maps of 500nm and 100nm areas of polystyrene-b-polybutadiene-b-polystyrene film showing surface locations with different mechanical properties. Brighter areas as one marked with a green cross are characterized by AvZ curve shown below (a) and common for polystyrene; darker areas as one marked with a blue cross—by AvZ curve shown below (b) and common for polybutadiene.

Measurements of AvZ curves, which are less destructive and are characterized by smaller tip-sample area, are more suitable for probing surface mechanical properties with unprecedented resolution of 1nm. This is demonstrated in Figures 6.13a–b by force volume maps, which are built of 128x128 arrays of AvZ curves collected on the surface of a thin film of polystyrene-block-polybutadiene-blockpolystyrene (SBS) triblock copolymer.

The curves representing the tip interactions with polystyrene and polybutadiene blocks are different (Figures 6.13a- b); the different contrasts in the maps reflect the variations in Z-travel needed to damp the amplitude of the interacting probe to a trigger level chosen by an operator. Such experiments allow precise identification of surface locations occupied by different blocks and offer experimental data for nanomechanical models for extracting quantitative data, which is a challenging problem currently under careful examination. In further expansion of nanomechanical characterization of polymers with AFM, it will be crucial to examine viscoelastic responses of different polymer systems in a broad frequency range and at various temperatures.

AFM studies of polymer materials can be performed not only at ambient temperatures but also at elevated temperatures, as shown in the examples above (Figures 6.10–11). Recently, studies were expanded towards low temperatures: an example of imaging of a multicomponent polymer material is presented in Figures 6.14a–c. These images were obtained on a sample of roofing material at different temperatures. The contrast of the phase image at T = -10°C differentiates a presumably rubber matrix and harder domains of another polymer, which have specific morphologies. The phase contrast has vanished at T = -35°C, yet the domains became visible in the height image, most likely, due to differences in thermal contraction of the components.

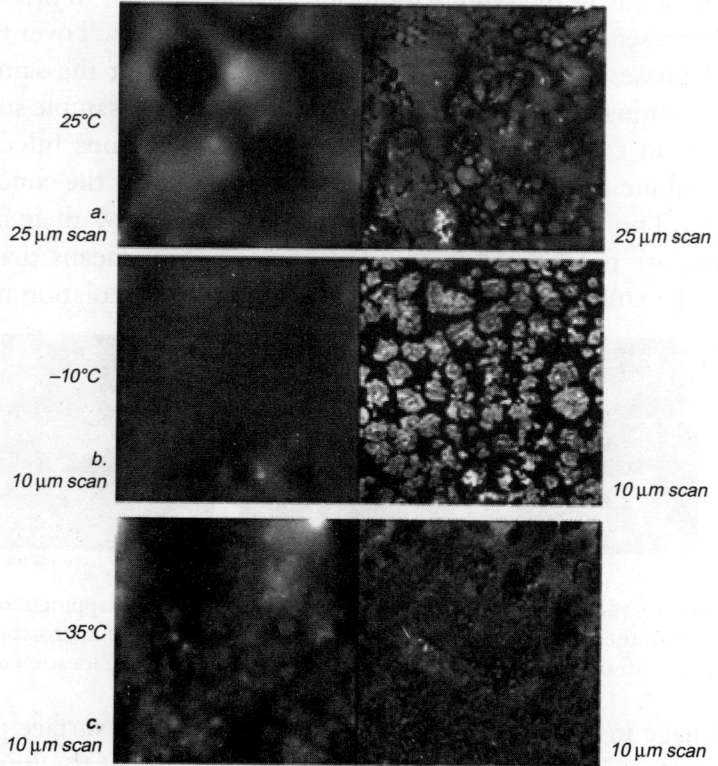

Fig. 6.14: Height (left) and phase (right) images of roofing material at different temperatures: (a) 25°C, (b) (-10°C) and (c) (-35°C).

The variations in mechanical properties of different components of the roofing material with temperature are reflected in these images. This helps characterize the sample morphology. An earlier study, which was performed on a series of block copolymers, demonstrated that differentiation of the components of a heterogeneous polymer material is best achieved when imaging is conducted at temperatures above the glass transition (T_g) of one component and below T_g of the other component.

With the increasing penetration of polymer materials into the semiconductor, data storage, and flat panel industries, the examination of conducting and semiconducting properties of these materials at small scales is becoming an important research area. The capabilities of AFM-related techniques such as electric force microscopy (EFM), surface potential microscopy, and conducting and tunneling AFM have yet to be fully explored. As an application example of these methods we consider the images of a polymer blends filled with carbon black (CB)—thermoplastic vulcanizate, which are shown in Figures 6.15a-c. The sample for this study was prepared with a cryo-ultramicrotome. The height image in Figure 6.15a reveals the surface topography of one of the sample locations. A large number of bright particles most likely present the CB particles, which are distributed all over the sample. The EFM phase image in Figure 6.15b was obtained at the same location when the scanning probe was scanning slightly above the sample surface. The darker spots in this image indicate the conducting regions filled with CB, which contribute to a percolation network responsible for the conductivity of this sample. The CB particles, which are recognizable by their dimensions (40-60 nm), are not limited to the dark regions. This means that merely a portion of the embedded particles contributes to the percolation network.

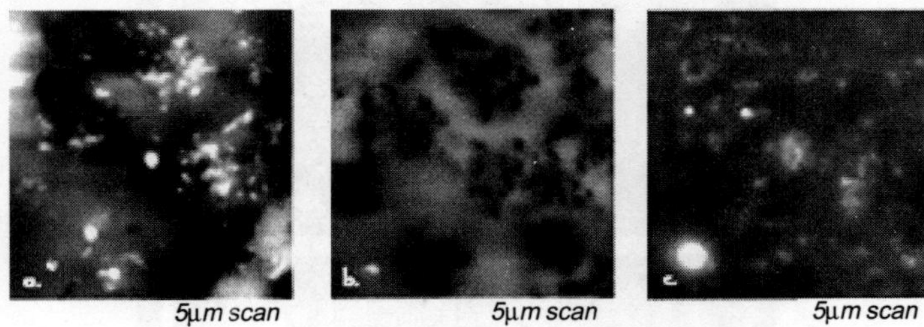

Fig. 6.15: (a) Height image of thermoplastic vulcanizate obtained in tapping mode. (b) Phase image of the same location as in (a) obtained in electric force microscopy mode with a probe lift of 20 nm. (c) Surface potential image of the same location as in (a).

The image in Figure 6.15c shows the distribution of surface potential at the same surface location. Bright spots (in the left side of the image) can be assigned to surface changes. Each of the images provides complementary information that can be invaluable in the characterization of electric properties of materials and their correlation with morphology and structure.

6.8 Polymer Science and Engineering

Polymer research in the School of Materials has a long and esteemed history. Presently, sixteen academic staff have polymer-related research interests and these cover a diverse range of polymer materials and research topics.

The focus is on fundamental studies of structure-property relationships for polymer materials of current and future importance, including controlled synthesis and processing, and effects of detailed chemical structure and nano-, meso- and macro-scale morphology on biological, chemical and physical properties The research is underpinned by a full range of modern analysis and characterisation facilities that are described below the following list of polymer research themes.

Research Themes

The polymer research activities are closely integrated and, for ease of reference, are grouped below into (mainly materials-based) themes, several of which are interrelated and encompass the research of different members of academic staff, both individually and in collaboration.

Many polymer research projects span several of these themes and topics.

- Biopolymers
- Coatings and Films
- Composites
- Deformation Micromechanics
- Fibres
- Multiphase Polymers
- Polymer Colloids
- Polymerisation
- Polymer Processing
- Responsive Polymers

Facilities

Polymer research is supported by a full suite of modern facilities that include:

- Several polymer synthesis laboratories.
- Gas-liquid and liquid-liquid chromatography for analysis of precursor materials and other small molecules.
- Gel permeation chromatography, viscometry, vapour pressure osmometry,

light scattering, UV-visible spectroscopy, infrared spectroscopy, Raman spectroscopyand X-ray diffraction for monitoring polymerisations and determination of molecular properties and structure, with access to nuclear magnetic resonance and mass spectroscopy facilities in the School of Chemistry.

- ❖ Differential scanning calorimetry (DSC), micro-DSC, dynamic mechanical analysis and thermogravimetry for thermal characterisation.
- ❖ Photon correlation spectroscopy, disc centrifugation and electrosonic electrophoresis for particle characterisation.
- ❖ Several optical, transmission electron and scanning electron microscopes for morphological characterisation.
- ❖ Infrared and Raman microscopes for spectroscopic investigation of surfaces, fibres and molecular deformation.
- ❖ Rheometers for rheological characterisation of polymer solutions, dispersions and melts.
- ❖ Injection moulding, extrusion, mixing, resin transfer moulding and rapid, reactive processing facilities for processing of polymers and composites.
- ❖ A wide range of mechanical testing equipment for determination of mechanical and failure behaviour, including a humidity and temperature controlled room dedicated to mechanical testing of polymers.

7

Biomedical Applications of Polymers

7.1 Biomedical Polymers

Introduction

Polymer chemistry is one of the most exiting topics in science because it goes beyond the preparation. It is multidisciplinary and gives direct access to the application areas. Within this broad field applications are biomedical polymers particularly exciting. Firstly, biomedical polymers require more than in other area that multidisciplinary approach. This means that cooperation with biomedical scientists, biologists, biomedical engineers, surgeons, etc. is mandatory, which is extremely inspiring. Secondly, it is highly motivating to make materials that can contribute to the quality of life of mankind. Finally, working on topics which are on the edge of common interest of several scientific disciplines will enhance the likelihood to discover breakthroughs. Polymers should be designed and developed to be used for certain specific applications, and thus the application must be an integral part of this science.

To develop properties for certain specified application properly it is mandatory to have an outstanding control of the chemistry. The cope with this in a practical way it is challenging to developed generic tools, generating a broad range of properties with a limited number of synthetic steps. Most studies of polymer science nowadays make use of a controlled polymerization of existing monomers. This is a very valuable way to prepare new structures and properties, but it is certainly not enough.

To extend the scope and, particularly, to enable the introduction of functionalities, we should go beyond this. We develop a concept to design and make a family of reactive compounds, provided with functional groups (Figure 7.1).

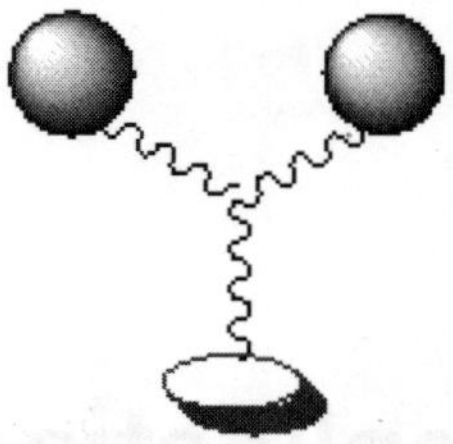

Fig. 7.1: Compound with reactive and functional groups.

More in detail, we developed a route to make polyurethanes based on a novel family if AB_2 monomers, in which many functional groups can be introduced. The requirements for biomedical applications are so complex that a polymer has to fulfill a number of functionalities simultaneously.

Biomedical polymers should fulfill at least three requirements:

1. The polymer should have the required mechanical properties.
2. The polymer should be biocompatible.
3. The polymer should have special functionalities to fulfill the specific requirements to do the desired task.

Since polyurethanes can more or less fulfill the first two requirements the focus of our work will be on the third requirement. We believe that hyperbranched polyurethanes based on our new family of AB_2 monomers are very suitable for that purpose.

Example

Many properties in biomedical applications are to a large extent determined by the surface characteristics of a device. Thus, control of the surface topology and chemistry is a key issue. One of the research topics is to develop coatings with a high number of reactive groups, to fix various functionalities simultaneously (figure below).

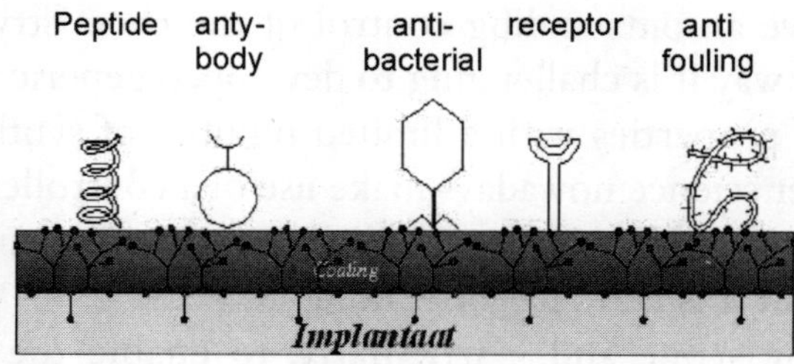

Thanks to the AB_2 monomers is the surface covered with reactive (blocked) isocyanate groups. Due to the high functionality can the surface be decorated with a number of biomedical functionalities. One advantage for the coating approach is that it is applicable on many implants.

Future Work in Brief

Control of the surface properties will be an important topic of our programs, whereas the focus will be on biomedical applications. Our new technology will be used to make hyperbranched polymers, provided with hydrophilic polymer tails. This will lead to anti-fouling properties, which is extremely important in implants. Another important topic is to introduce antibacterial properties in implants. One of the main mode of failure with implants is infections. The high functionality of the hyperbranched polymers is also an enabling technology to introduce antifouling and antibacterial properties at the same time. A last topic to mention here is tissue engineering. In tissue engineering the surface have to be covered with peptide unites as anchor points for human cells. The high functionality offers a way to achieve that goal.

Vacancy

Master students with a background in organic/polymer chemistry are invited to contact me for a more detailed discussion on one of the topics that is briefly mentioned above.

Background of Ton Loontjens

Ton Loontjens has very recently been appointed as part time professor at the university of Groningen. Next to this he holds a position as principal scientist at DSM research in Geleen, the Netherlands. After his study of organic chemistry at the university of Nijmegen he started in 1975 at DSM as group leader of the polypropylene polymerization group. A few years later he was also responsible for the polyethylene polymerization group. In 1985 he was appointed as head of a polymer chemistry department on coating resins, melamine resins and stabilization of polymers. In 1992 he became head of another polymer chemistry department on polyesters, polyamides and coating resins. Since 1995 he has held the position of principal scientist.

7.2 Contract Research

Formulation support and problem solving fall under our contract research service. Comprehensive understanding of physicochemical properties can be used to help successful acceleration through regulatory submissions and aid reduction in late stage failures. To facilitate this we provide a range of bespoke solid state analyses for drug substance and formulation issues enabling prediction, problem solving and optimisation earlier in the R&D process.

Application Area

The following is a list of links to selected application areas:

- ❖ API & excipient,
- ❖ Solid dosage,
- ❖ Controlled release,
- ❖ Inhalation,
- ❖ Biomedical polymers, and
- ❖ Biopharmaceutical.

7.3 Degradable Polymers for Biomedical and Biomaterial Applications from Sigma Aldrich

Background

During the last two decades, significant advances have been made in the development of biocompatible and biodegradable materials for biomedical applications, and in the case of the latter category, industrial applications, as well. In the biomedical field, the goal is to develop and characterize artificial materials or, in other words, "spare parts" for use in the human body to measure, restore, and improve physiologic function, and enhance survival and quality of life. Typically, inorganic (metals, ceramics, and glasses) and polymeric (synthetic and natural) materials have been used for such items as artificial heart-valves, (polymeric or carbon-based), synthetic blood-vessels, artificial hips (metallic or ceramic), medical adhesives, sutures, dental composites, and polymers for controlled slow drug delivery. The development of new biocompatible materials includes considerations that go beyond nontoxicity to bioactivity as it relates to interacting with and, in time, being integrated into the biological environment as well as other tailored properties depending on the specific "in vivo" application.

Biomimetics

The parallel field of "biomimetics" may be described as the "abstraction of good design from nature" or, plainly put, the "stealing of ideas from nature". The goal is to make materials for non-biological uses under inspiration from the natural world by combining them with manmade, non-biological devices or processes. This is fast becoming a new research frontier.

Biocompatible Polymers

One area of intense research activity has been the use of biocompatible polymers for controlled drug delivery. It has evolved from the need for prolonged and better control of drug administration. The goal of the controlled release devices is to maintain the drug in the desired therapeutic range with just a single dose. Localized delivery of the drug to a particular body compartment lowers the systemic drug level, reduces the need for follow-up care, preserves medications that are rapidly destroyed by the body, and increases patient comfort and/or improves compliance. In general, release rates are determined by the design of the system and are nearly independent of environmental conditions.

Degradable Polymers

A variety of natural, synthetic, and biosynthetic polymers are bio- and environmentally degradable. A polymer based on the C-C backbone tends to be nonbiodegradable, whereas heteroatom-containing polymer backbones confer biodegradability. Biodegradability can therefore be engineered into polymers by the judicious addition of chemical linkages such as anhydride, ester, or amide bonds, among others. Figure 7.2 provides a schematic representation of the types of polymer degradation. The mechanism for degradation is by hydrolysis or enzymatic cleavage resulting in a scission of the polymer backbone. Macroorganisms can eat and, sometimes, digest polymers, and also initiate a mechanical, chemical, or enzymatic aging.

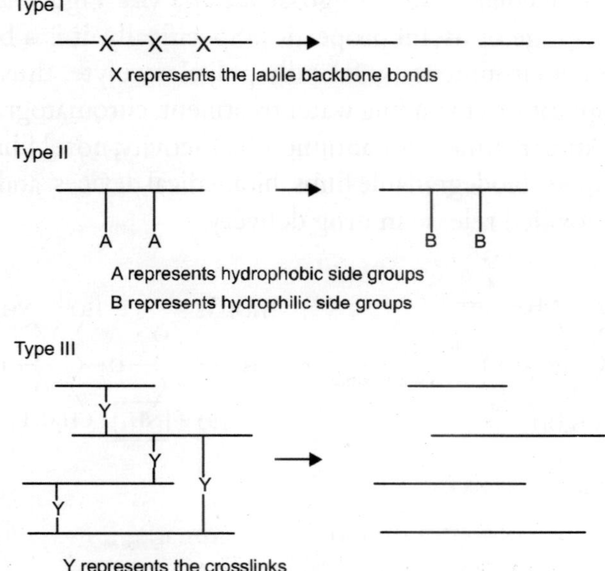

Fig. 7.2: Schematic representation of the types of polymer degradation.

Properties of Degradable Polymers that Make Them Suitable as Biomaterials

Biodegradable polymers with hydrolyzable chemical bonds are being researched extensively for biomedical, pharmaceutical, agricultural, and packaging applications. In order to be used in medical devices and controlled-drug-release applications, the biodegradable polymer must be biocompatible and meet other criteria to be qualified as a biomaterial-processable, sterilizable, and capable of controlled stability or degradation in response to biological conditions. The degradation products often define the biocompatibility of a polymer, not necessarily the polymer itself. Poly(esters) based on polylactide (PLA), polyglycolide (PGA), polycaprolactone (PCL), and their copolymers have been extensively employed as biomaterials. Degradation of these materials yields the corresponding hydroxy acids, making them safe for in vivo use.

Other Bio/Environmentally Degradable Polymers

Other bio/environmentally degradable polymers include poly(hydroxyalkanoate)s of the PHB-PHV class, additional poly(ester)s, and natural polymers, particularly, modified poly(saccharide)s, e.g., starch, cellulose, and chitosan.

Chitosan

Chitosan is a technologically important biomaterial. Chitin is the second most abundant natural polymer in the world after cellulose. Upon deacetylation, it yields the novel biomaterial Chitosan, which upon further hydrolysis yields an extremely low molecular weight oligosaccharide (*see* Figure 7.3). Chitosan possesses a wide range of useful properties. Specifically, it is a biocompatible, antibacterial and environmentally friendly polyelectrolyte, thus lending itself to a variety of applications including water treatment, chromatography, additives for cosmetics, textile treatment for antimicrobial activity, novel fibres for textiles, photographic papers, biodegradable films, biomedical devices, and microcapsule implants for controlled release in drug delivery.

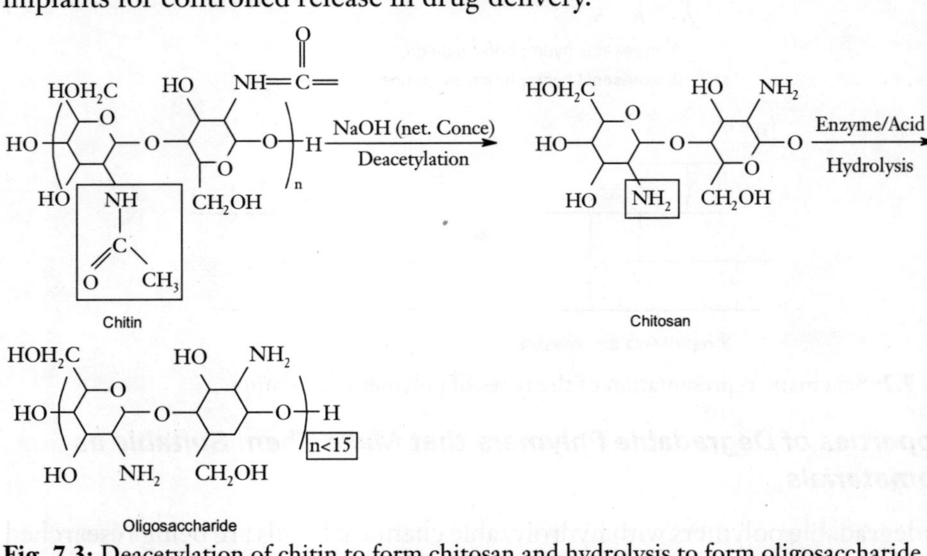

Fig. 7.3: Deacetylation of chitin to form chitosan and hydrolysis to form oligosaccharide.

Poly(ethylene Oxide)

Poly(ethylene oxide), PEO, a polymer with the repeat structural unit -CH_2CH_2O-, has applications in drug delivery. The material known as poly(ethylene glycol), PEG, is in fact PEO but has in addition hydroxyl groups at each end of the molecule. In contrast to high molecular weight PEO, in which the degree of polymerization, n, might range from 10^3 to 10^5, the range used most frequently for biomaterials is generally from 12 to 200, that is PEG 600 to PEG 9000, though grades up to 20,000 are commercially available. Key properties that make poly(ethylene oxide) attractive as a biomaterial are biocompatibility, hydrophilicity, and versatility. The simple, water-soluble, linear polymer can be modified by chemical interaction to form water-insoluble but water-swellable hydrogels retaining the desirable properties associated with the ethylene oxide part of the structure. Poly(ethylene glycol)s first appeared in the U.S. Pharmacopoeia in 1950. Since then they have been used increasingly for a variety of pharmaceutical applications.

Multiblock Copolymers of Poly(ethylene Oxide) and Poly(butylene Terephthalate)

Multiblock copolymers of poly(ethylene oxide) (PEO) and poly(butylene terephthalate) (PBT) are also under development as prosthetic devices and artificial skin and as scaffolds for tissue engineering. These materials are subject to both hydrolysis (via ester bonds) and oxidation (via ether bonds). Degradation rate is influenced by PEO molecular weight and content. Additionally, the copolymer with the highest water uptake degrades most rapidly.

A widely used nondegradable polymer is ethylene-vinyl acetate copolymer. This copolymer displays excellent biocompatibility, physical stability, biological inertness, and processability. In drug delivery application these copolymers usually contain 30-50 weight percent vinyl acetate. Ethylene-vinyl acetate copolymer membrane acts as the rate-limiting barrier for the diffusion of the drug. In the Type II class of degradable polymers, the conversion of the hydrophobic substituents to hydrophilic side groups is a first step in the degradation process. A team of researchers has addressed the problem of fabricating open-pore, biodegradable polymer scaffolds for cell seeding or other tissue engineering applications. The material selected was the tyrosine-derived polycarbonate poly(DTE-co-DT carbonate), in which the pendant group via the tyrosine, an amino acid, is either an ethyl ester (DTE) or free carboxylate (DT). Through alteration of the ratio of DTE to DT, the material's hydrophobic/hydrophilic balance and rate of in vivo degradation can be manipulated. It was shown that, as DT content increases, pore size decreases, the polymers become more hydrophilic and anionic, and cells attach more readily.

Water Swellable Polymer Networks

Water-swellable polymer networks may function as hydrogels at one end or as superabsorbers at the other extreme. Hydrogels are characterized by the pronounced affinity of their chemical structures for aqueous solutions in which they swell rather than dissolve. Such polymeric networks may range from being mildly absorbing, typically retaining 30 wt. % of water within their structure, to superabsorbing, where they retain many times their weight of aqueous fluids. Several synthetic strategies have been proposed to prepare absorbent polymers including:

- ❖ Polyelectrolyte(s) subjected to covalent cross-linking.
- ❖ Associative polymers consisting of hydrophilic and hydrophobic components ("effective" cross-links through hydrogen bonding).
- ❖ Physically interpenetrating polymer networks yielding absorbent polymers of high mechanical strength.

Clearly, these strategies are not mutually exclusive, and efforts have focused on tailoring composite gels which are critically reliant on the balance between polymer-polymer and polymer-solvent interactions under various stimuli including changes in temperature, pH, ionic strength, solvent, concentration, pressure, stress, light intensity, and electric or magnetic fields. Such stimuli-responsive polymers, the so-called smart gels, continue to be the subject of extensive investigation for applications in diverse fields. These applications range from biomedical (controlled drug release, ocular devices, and biomimetics), agricultural (soil additive to conserve water, plant root coating to increase water availability, and seed coating to increase germination rates), and personal care (diapers and adult hygiene products), to industrial (thickener, gelling agent, cable wrap, specialty packaging, tack reduction for natural rubber, and fine coal dewatering).

Absorbent polymers may be of synthetic (petrochemical) origin where the effects of morphology and porosity affects the absorbent properties. Aldrich also offers an extensive selection of polymers of natural (starches, etc.) and semisynthetic (cellulose ethers, etc.) origins for use in the synthesis of multicomponent hydrogels. To aid in designing your application-specific hydrogel, Aldrich offers over 1,500 monomers and a wide selection of cross-linking agents.

7.4 Biomedical Applications of Polymers

The PI has participated in many polymer studies involved biomedical applications. Through an grant from US National Institute of Health the PI has worked on immobilized enzymes on polyacrylamide. Also through several industry projects the PI worked on surface modification of catheter tubings.

Polymers can be used in many areas of biomedical applications: Artificial skins, hip replacement, controlled release, blood bags, heart valves, intraocular lens, and breast implant. Each of these applications required a unique combination of properties. The diverse nature of polymers offers the opportunity to match their properties for the applications.

7.5 Polymers for Biomedical Applications

We have capabilities for designing and synthesising polymers for use in medical implants, and tissue engineered products and therapies:

- Our expertise,
- Current uses,
- Outcomes, and
- Our partners.

Our Expertise

We design novel polymers with mechanical properties and biological performance appropriate to the desired-end products in ophthalmic and tissue engineering fields.

Current Uses

CSIRO is using this capability to develop suitable polymers for an implantable contact lens and injectable polymer gel for restoring accommodation. The implantable contact lens will enable safe, permanent, but reversible, refractive correction.

We are also developing synthetic and recombinant polymers for tissue engineering and wound healing applications.

Outcomes

Research with the Cooperative Research Centre (CRC) for Cardiac Technology and its partners led to the development of a family of biostable polyurethanes (Elast-Eon™) that has exceptional biostability, while maintaining excellent flexibility. Elast-Eon™ provides, for the first time, a material that has an ideal balance between biostability and flexibility.

This technology led to the establishment of AorTech Biomaterials, a company with a current focus on commercialising Elast-Eon™ polyurethanes for medical implants such as heart valves, breast implants and pace maker leads. Our development of new materials for ophthalmology is in collaboration with the CRC for Eye Research and Technology. The CIBA vision extended wear contact lens (Focus Night & Day) is a result of designing revolutionary materials capable of transmitting up to six times more oxygen to the eye than ordinary lenses.

We have developed and patented a family of biostable porous in situ curable fluoropolymer gels. The target application is to replace the conventional intraocular lens market with a new product that restores the ability of the eye to focus at a range of distances (presbyopia).

We have also developed a family of biodegradable polymers that now forms the basis of the spin-out companyt PolyNovo Biomaterials Pty Ltd.

Our Partners

Our work is supported by a number of strong interactions with collaborators who bring skills for in vivo evaluation, and prototype device design and fabrication.

7.6 Biomedical Polymers

Reference		*DHBIOM00000017*
Taught in		Majors List Master of Biomedical SciencesMajors List Master of Biomedical Sciences
Theory	(A)	25.0
Exercises	(B)	0.0
Training and projects	(C)	20.0
Studytime	(D)	180.0
Studypoints	(E)	6
Level		
Credit contract?		Access is determined after successful competences assessment
Examination contract?		This course can not be taken through this kind of contract
Credit contract mandatory if Exam contract?		Separate credit contract mandatory
Retake possible?		Yes
Teaching Language		Dutch
Lecturer		Peter Dubruel
Department		WE07
Co-lecturers		Etienne Schacht

Position of the Course

The student gathers knowledge and insight of the synthesis, characterisation and shaping of polymeric materials for biomedical applications in particular those materials with a mechanical function and a therapeutical function. In addition the student becomes acquainted with the use of polymers in innovative biomedical applications.

Contents

Review of the basic principles of the synthesis of polymers: vinyl polymers, addition and condensation polymers, polymerisation by ring opening, biodegradable polymers, hydrogels, chemical modification of polymers.Properties of polymers: bulk and surface characterisation, effect of sterilisation.Shaping techniques.Applications of polymers in biomaterials and artificial organs. Applications of polymers as scaffolds for tissue engineering.

Starting Competences

Having successfully completed the courses Organic chemistry and Human pathogenesis from the bachelor program biomedical sciences, or having acquired the relevant ending objectives by other means.

Final Competences

- Knowing the basic principles of the synthesis of polymers for biomedical applications.
- Having a thorough knowledge on the properties of polymers for biomedical use.
- Being able to determine the properties of biomedical polymers for specific biomedical purposes.
- Knowing the design of polymers.
- Being able to indicate the usefulness of polymers for innovative biomedical applications.

Teaching and Learning Material

Syllabus and in additon recent actual articles.

7.7 Development of Piezoelectric Polymer for Biomedical Applications

Research Interests

Development of piezoelectric polymer for biomedical applications. Piezoelectric polymers such as PVDF copolymer have been investigate for use in biomedical applications such as sensors and actuators because of it biocompatibility, flexibility, and easy processing.

In this study, we investigate the reverse effect of piezoelectric polymer and observe how this stimulation affects cell growth. The schematic diagram in Figure 7.4, showed a small displacement being produced by applying an alternating electrical field to the piezoelectric polymers.

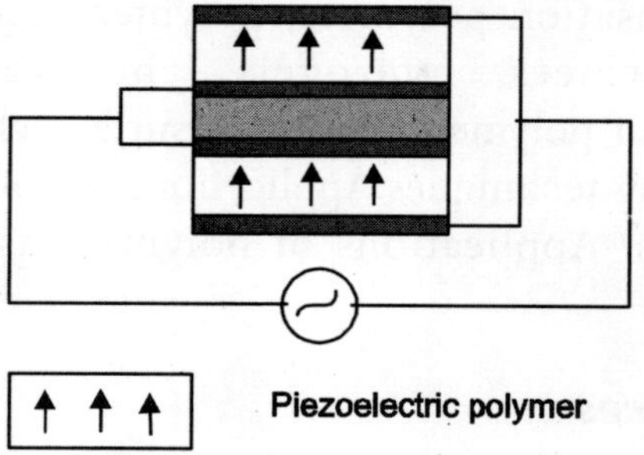

Fig. 7.4: Piezoelectric polymer under electric field.

Under a sterile environment, the number of cells initially seeded onto the piezoelectric polymer was being study, whereby such displacement would affect cell behavior (proliferation, adhesion, spreading or differentiation), as shown in Figure 7.5. One of the objectives is also to explore the seeding condition under static and non-static seeding system. Our goal is to fabricate an actuating biopolymer for biomedical applications.

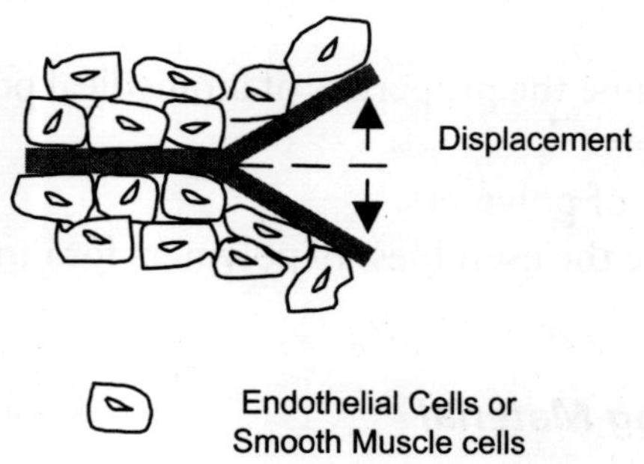

Fig. 7.5: Displacement effects on cell behavior.

Our preliminary results under static environment had indicated a positive endothelialization behavior on PVDF-TrFE membranes. Light micrographs of endothelial cells at various stages of development, was evident as in Figure 7.7. The viability and physiological function of endothelial cells that grow on PVDF-TrFE membranes, are thus being studied further.

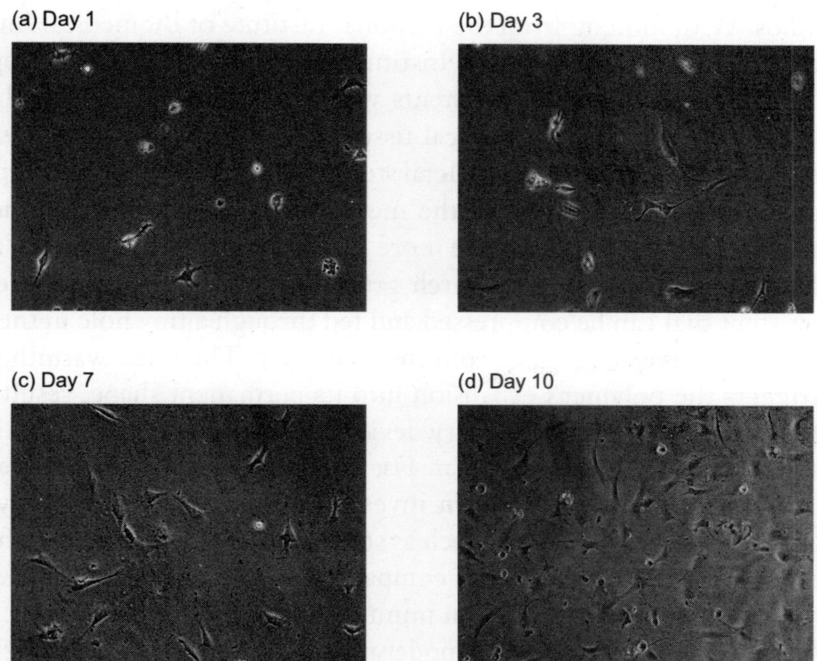

Fig 7.7: Endothelial cells behavior on PVDF-TrFE membranes. (Magnification 10x).

7.8 Biomedical Applications For Shape-Memory Polymers

Researchers at the Georgia Institute of Technology are developing unique polymers, which change shape upon heating, to open blocked arteries, probe neurons in the brain and engineer a tougher spine.

These so-called shape-memory polymers can be temporarily stretched or compressed into forms several times larger or smaller than their final shape. Then heat, light or the local chemical environment triggers a transformation into their permanent shape. "My focus has been to optimize these polymers for many different biomedical applications. My lab studies how altering the chemistry and structure of the polymers affects their chemical, biological and mechanical properties," said Ken Gall, a professor in the George W. Woodruff School of Mechanical Engineering and School of Materials Science and Engineering. The mechanical properties of these polymers make them extremely attractive for many biomedical applications, according to Gall, who described his research in this area during two presentations at the Materials Research Society's fall meeting in November.

Engineers are always searching for materials that display unconventional properties able to satisfy the severe requirements for implantation in the body. Particular attention must be paid to the biofunctionality, biostability and biocompatibility of these materials, which come into contact with tissue and body fluids. With funding from the National Institute of Biomedical Imaging and Bioengineering of the National Institutes of Health (NIH), Gall proposed replacing metallic cardiovascular stents with plastic ones because polymers more closely resemble soft biological tissue. Plus, polymers can be designed to gradually dissolve in the body. "Metal stents are frequently covered in plastic anyway, so we set out to remove the metal leaving just a polymer sheath," explained Gall. "Also, polymers are more flexible and do not stress the artery walls like the metals." Gall's research group has designed a shape-memory polymer stent that can be compressed and fed through a tiny hole in the body into a blocked artery, just like a conventional stent. Then, the warmth of the body triggers the polymer's expansion into its permanent shape, resulting in natural deployment without auxiliary devices. This work was published in the journal Biomaterials earlier this year. For another project, Gall and graduate student David Safranski have been investigating how altering a polymer's chemistry changes its properties, such as stretchiness. This project was funded by MedShape Solutions, an Atlanta company that Gall co-founded to develop medical devices primarily for use in minimally invasive surgery.

"You can tailor the polymer to moderate its strength, stiffness, stretchiness and expansion rate," noted Gall.

They found that by changing the chemistry of the polymer backbone to include special side groups, they could increase of the amount of strain the polymer could withstand before failing without sacrificing stiffness. This discovery enabled the creation of polymers that could stretch farther and also push harder during recovery.

Gall and graduate student Scott Kasprzak are exploring how these polymers might be used as a deployable neuronal probe, with funding from the National Institute of Neurological Disorders and Stroke of the NIH.

"We're looking for smart materials that can be synthesized in the size range of 100 microns—similar to the size of a strand of hair—and then be inserted into brain tissue," explained Gall. "This type of probe would need to slowly change shape inside the brain as to not disturb any surrounding tissue."

Another project in Gall's laboratory is examining the use of these polymers for the spine. Most spinal surgeries are currently not performed arthroscopically, so Gall sees benefits in using these shape-memory materials to enable minimally invasive spinal surgery.

With funding from the National Institute of Arthritis and Musculoskeletal and Skin Diseases (NIAMS), Gall and graduate student Kathryn Smith are developing shape-memory polymers for the spine that are tough—meaning they stretch far and support a lot of weight like native spinal disks.

"This would improve the deliverability and life of artificial disks currently used in the spine. Essentially, we're just trying to engineer tougher synthetic polymers that can be easily delivered," explained Gall, who is collaborating on this project with Barbara Boyan and Johnna Temenoff, both of the Coulter Department of Biomedical Engineering at Georgia Tech and Emory University.

In addition to exploring different biomedical applications for shape-memory polymers, Gall has also turned his attention to manufacturing them. Walter Voit, a graduate student in the Technological Innovation: Generating Economic Results (TI: GER) program, is investigating how to produce shape-memory polymers at a low cost. More specifically, Voit is examining different types of materials and processing methods that can be used to commercially produce quality polymers for lower cost medical applications.

8

Macromolecules of Interfaces and Films

8.1 Macromolecules: Natural and Synthetic Polymers

Objectives

In this laboratory you will become familiar with the classifications of polymers by synthesizing and examining several of the following:

1. A linear condensation copolymer (GlyptalTM resin),
2. A branched addition polymer (polymethylmethacrylate),
3. A cross-link a natural linear polymer (cellulose),
4. A loosely cross-linked silicon-based condensation polymer (a polymethylsiloxane), and
5. A cross-linked polyvinyl alcohol .

Introduction

Approximately 50 per cent of the industrial chemists in the United States work in some area of polymer chemistry, a fact that illustrates just how important polymers are to our economy and standard of living. These polymers are essential to the production of goods ranging from toys to roofing materials. So what exactly are polymers? Polymers are substances composed of extremely large molecules termed macromolecules, with molecular masses ranging from 104 to 108 amu. The macromolecules consist of many smaller molecular units, monomers joined together through covalent bonds. The molar mass of the polymer is quoted as an average molar mass.

Both natural and synthetic polymers are ubiquitous in our lives: elastomers (polymers with elastic, rubber-like properties), plastics (the first plastic was used in 1843 to make buttons), textile fibers, resins, and adhesives. The more common polymers include acrylics, alkyds, cellulosics, epoxy resins, phenolics, polycarbonates, polyamides, polyesters, polyfluorocarbons, polyolefins, polystyrenes, silicones, and vinyl plastics, to name but a few.

Naturally occurring macromolecules are obviously derived from living things ñ wood, wool, paper, cotton, starch, silk, rubber ñ and have provided us for centuries with materials for clothing, food, and housing. Starch, glycogen, and cellulose are all polymeric versions of the monomer glucose. Proteins are macromolecules composed of monomeric units of alpha amino acids; nucleic acids are composed of subunits (nucleotides) containing a nitrogeneous base, sugar and phosphate groups. Natural rubber is a latex exudate of certain trees and composed of monomers called isoprene units. The usefulness of latex was first discovered by Lord Mackintosh in Malayasia in the last century and provided the foundation of his waterproof rainwear empire.

The temptation to improve upon nature has always been great and has rarely been resisted. When scientists linked the special properties of these substances (physical properties such as tensile strength and flexibility) to the sizes of their molecules the next logical step involved chemical modifications of naturally occurring polymers.

Synthetic celluloid derives from natural cellulose and stems from an accident that Christian Schoenbein, a chemistry professor had in 1846—the age of plastic had begun although, initially the interest in cellulose nitrate was more for their explosive properties. When cellulose (from wood chips or fiber) is treated with a mixture of nitric acid, camphor, and alcohol, the resultant product is called CelluloidTM and bears very little resemblance to the starting material. CelluloidTM possess the ability to be molded into hard, smooth billiard balls (replacing the original, very expensive ivory balls) and into thin sheets for making movie pictures. CelluloidTM is highly flammable and today has been replaced by greatly improved synthetic polymers such as bakelite discovered in 1907 by the Belgian-American Chemist, Leo H. Baekeland. When cellulose is treated with sodium hydroxide and carbon disulfide (CS2), cellulose xanthate is formed. A viscous (thick) solution of cellulose.xanthate, forced through fine holes into dilute sulfuric acid, regenerates the cellulose as fine, continuous cylindrical threads called rayon. If the solution is forced instead through a narrow slit, a thin transparent film or sheet is obtained called cellophane.

Thermoplasts Versus Thermosetting

Polymers generally are classified into two broad groups in accordance with their behavior upon heating. Polymers that can be repeatedly melted and solidified (without damage) are said to be thermoplasts; those that solidify once but will not melt again with damage are said to be thermosets. Technically, though only thermoplasts are true plastics, even though the term "plastic" is commonly applied to all synthetic polymers.

The thermoplastic substance contains long, thin molecules which form tangled chains and is rigid at lower temperatures but gradually softens upon the application of heat and after it passes a characteristic temperature known as its glass transition temperature (Tg). Below Tg, the substance is brittle, having the characteristic properties of a glass; above Tg, the substance becomes flexible and soft. Chewing gum is a thermoplast that becomes extremely brittle when the outside temperatures drop below its glass transition temperature—this is an useful property to use in order to remove chewing gum from your clothes. Once warmed above Tg, however, the gum quickly softens and regains its flexibility. Some thermoplasts, such as polystyrene, melt before reaching their glass transition temperatures and remain rigid materials up to their melting points. Thermoplastic polymers are used frequently for injection molding of such items as food storage containers and toys that are not exposed to high temperatures. Additionally, thermoplastic polymers can be molded, pressed and extruded.

Thermosetting substances contain large, cross-linked molecules and are also rigid at lower temperatures, undergo irreversible chemical and physical changes (including decomposition) upon heating. Such substances remain solids at higher temperatures than do thermoplastic materials, and they do not melt. Thermosets are often employed in high-temperature environments, such as for electrical insulation in electric motors and gasoline engines.

Thermoplastic polymers are composed of small monomers covalently bonded end-to-end in a long chain but without covalent bonds joining adjacent chains. Such macromolecules constitute the linear polymers. Shorter side groups attached to the long chains at periodic intervals, cause the polymers to be termed branched polymers. The chains, having average molecular masses up to one million amu, may be independent of each other (as in polyethylene) or loosely lined through hydrogen bonding (as in nylon). If the long chains are linked by covalent bonds, the polymeric network becomes two- or three-dimensional, resulting in an infusible (nonmelting) and insoluble material. Such macromolecules make up the cross-linked polymers, and are found in thermosetting materials. Polymers of all types in which the long chains are produced by joining two or more different kinds of monomers are termed copolymers.

The process by which the polymerization reaction occurs permits classification of polymers into two categories: addition and condensation polymers. Addition polymers

are those in which the monomers join at unsaturated carbon atoms; several are summarized in the Table. During polymerisation, the double bonds between the pairs of carbon atoms 'open up' and the carbon atoms of separate ethylene molecules join together to form a molecule of polyethylene. The first polymersiation of ethylene was accomplished in 1933 by the use of very high pressure (1000 atm) and oxygen as a catalyst. Nowadays, with the development of the use of powerful catalysts, addition can occur at atmospheric pressure. Polymethyl methacrylate, also called LuciteTM or PlexiglasTM (originally developed as an unbreakable substitute for glass in airplane canopies), belongs to this group of addition polymers. The polymerization is initiated by a variety of substances (such as benzoyl peroxide) that can form a free radical with the unsaturated carbon atom. The resulting addition polymer is described as a branched polymer.

The second way to make a polymer is by condensation polymerisation. In this process, two compounds with reactive atoms at the end of their molecules react together, usually with the release of a small molecular unit such as water or hydrogen chloride. The presence of two or more functional groups in the monomer usually leads to the production of a cross-linked polymer. GlyptalTM resin, formed by the reaction between phthalic acid and glycerol, is a condensation copolymer, a copolymer since 2 different types of monomers combine to form the chain.

Monomers are linked through ester ($-C(=O)-O-CH_2-$) bonds; the resin is termed a polyester. The more reactive phthalic anhydride is often used in place of phthalic acid in this reaction.

Paper is composed of naturally occurring cellulose, the polymeric structural material of plants. The cellulose chains are composed of linear glucose units. Parchment paper is made by cross-linking the linear cellulose polymer to form a sheet-like structure. Adjacent chains are cross-linked by ether ($-CH_2-O-CH_2-$) bonds resulting from dehydration of the alcohol groups (through the use of sulfuric acid as the dehydrating agent). Resulting reactions that form the macromolecules required for polymerization by either addition or condensation processes must be capable of proceeding indefinitely.

By far the most interesting uses of polymers involves replacement of diseased, worn out, or missing parts of the human body including flexible replacements for major blood vessels, replacement valves for hearts, temporary skin, and artificial joints. Artificial ball-and-socket pelvic (hip) joints made of steel (ball) and plastic (socket) are installed at the rate of 25,000 per year. People with crippling arthritis, debilitating coronary and circulatory problems, and burns all benefit from the development of biomedical polymers. Preventive and cosmetic dentistry, as well, benefit from polymers used to seal porous teeth and reconstruct missing enamel.

Linear silicones, or polysiloxanes, are comparatively new polymers based upon silicon-oxygen-silicon linkages. These polymers may be cross-linked to various degrees by additional -Si-O-Si- bonding between adjacent chains:

The R group is generally a hydrocarbon group such as $-CH_3$ (methyl), $-CH_2CH_3$ (ethyl), or $-C_6H_5$ (phenyl). Silicones are stable at much higher temperatures than carbon-based polymers, yet they remain flexible even at exceedingly low temperatures. Among such silicones is Silly PuttyTM, the cross-linked polymerization product of dimethyldichlorosilane, $(CH_3)_2Si(OH)_2$ with the release of hydrogen chloride (HCl):

$$Cl-\underset{CH_3}{\overset{CH_3}{Si}}-Cl + H_2O \longrightarrow HO-\underset{CH_3}{\overset{CH_3}{Si}}-OH + 2HCl$$

The unstable dimethyldihydroxysilane condenses rapidly to form a low molecular mass linear polymethylsiloxane polymer with the elimination of water:

$$Cl-\underset{CH_3}{\overset{CH_3}{Si}}-Cl \longrightarrow HO-\underset{CH_3}{\overset{CH_3}{Si}}-OH + (n-1)H_2O$$

This material is an oily liquid. Additional heating continues the polymerization to form longer chains. The addition of boron trioxide allows three such chains to be joined to form Silly PuttyTM:

O—Si—O—Si—
—Si—O—Si—O—B—
O—Si—O—Si—

This medium molecular mass polymer has physical properties between those of a fluid and an elastomer. Although it is resistant to rapid deformation, it flows easily with slowly applied stress.

Table 5.1: Addition Polymers.

Example	*Monomer(s)*	*Polymer*	*Use*
Polyethylene	$CH_2=CH_2$	$-CH_2-CH_2-$	Common polymer: bags, wire insulations, squeeze bottles
Polypropylene	$CH_2=CH(CH_3)$	$-CH_2-CH(CH_3)-$	Fibers, indoor-outdoor carpet, bottles, rope
Polystyrene	$NC(=CH_2)$–C_6H_5	$-H_2C-CH(C_6H_5)$	Styrofoam™; drinking cups, building insulation, packing materials
Polyvinylchloride (PVC)	$H_2C-CH(Cl)$	$H_2C-CH(Cl)-$	synthetic leathers, clear bottles, floor coverings, phonograph records, water pipes
Polytetrafluoroethylene (Teflon™)	$CF_2=CF_2$	–	nonstick surfaces, chemically resistant films, cookware coatings
Polymethylmethacrylate (Lucite™, plexiglass™)	$CH_2=C(CO_2CH_3)(CH_3)$	$-CH_2-C(CO_2CH_3)(CH_3)-$	unbreakable glass, latex paints
Polyacrylonitrile (Orlon™, Acrilan™, Creslan™)	$-H_2C=CH(CN)$	$-H_2C-CH(CN)-$	fibers for sweaters, blankets, carpets

(*Contd.*)

(*Contd.*)

Example	*Monomer(s)*	*Polymer*	*Use*
Polyvinylacetate (PVA)	$CH_2{=}CH_2$ with $-OCCH_3$ (C=O) substituent	$-CH_2-CH_2-$ with $-OCCH_3$ (C=O) substituent	adhesives, latex paints, chewing gum, textile coatings
natural rubber	$CH_2{=}C(CH_3)CH{=}CH_2$	$-CH_2{=}C(CH_3){=}CH-CH_2-$	The polymer is cross-linked with sulfur (vulcanization).
Polychloroprene (neoprene rubber)	$CH_2{=}C(Cl)CH{=}CH_2$	$-CH_2-C(Cl){=}CH-CH_2-$	cross-linked with zinc oxide; resistant to oil, gasoline
Styrene-Butadiene Rubber (SBR)	$C_6H_5-HC{=}CH_2$; $CH_2{=}CHCH{=}CH_2$	$H_2C-C(H)(C_6H_5)-CH_2-CH{=}CH-CH_2$	cross-linked with peroxides; most commonly used for tires, 25% styrene, 75% butadiene

Table 5.2: Condensation Polymers.

Example	*Monomer(s)*	*Polymer*	*Use*
Polyamides (nylon)	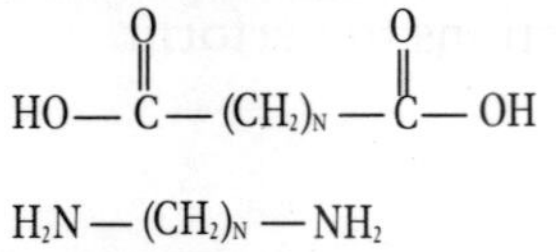 $HO-C(=O)-(CH_2)_N-C(=O)-OH$ $H_2N-(CH_2)_N-NH_2$	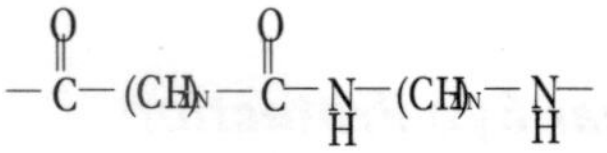 $-C(=O)-(CH_2)_N-C(=O)-NH-(CH_2)_N-NH-$	fibers, molded objects
Polyesters (Dacron™, Mylar™, Fortrel™)	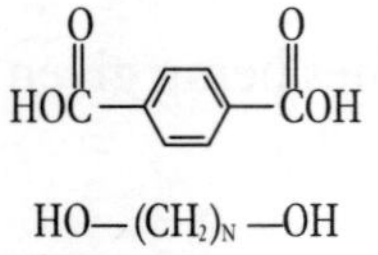 $HOC(=O)-C_6H_4-C(=O)OH$ $HO-(CH_2)_N-OH$	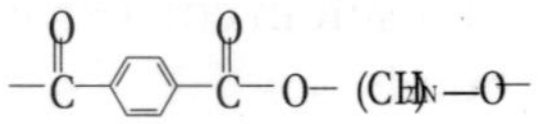 $-C(=O)-C_6H_4-C(=O)-O-(CH_2)_N-O-$	linear polyesters, fibers, recording tape
Polyesters (Glyptal™ resin)	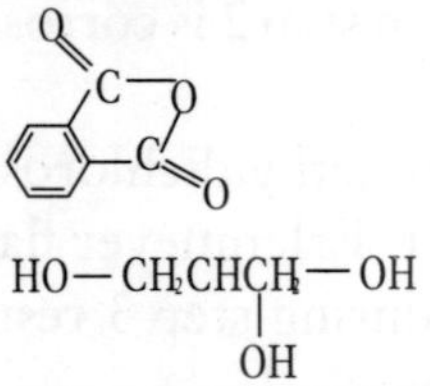 phthalic anhydride $HO-CH_2CHCH_2-OH$ (with OH on central carbon)	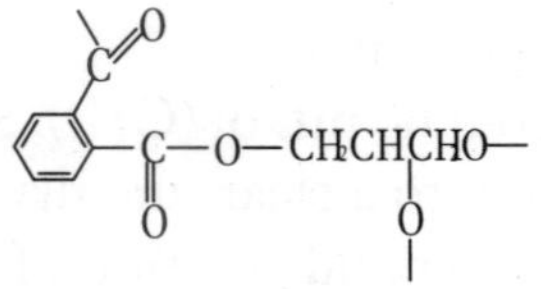 $-C(=O)-C_6H_4-C(=O)-O-CH_2CHCH_2O-$ (with O– on central carbon)	crosslinked polyester; paints
Polyesters (casting resin)	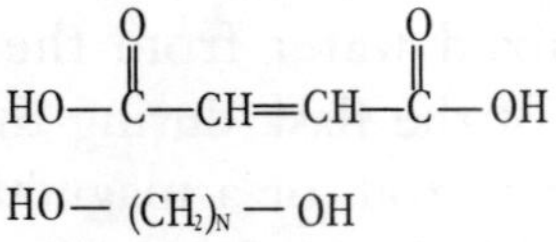 $HO-C(=O)-CH=CH-C(=O)-OH$ $HO-(CH_2)_N-OH$	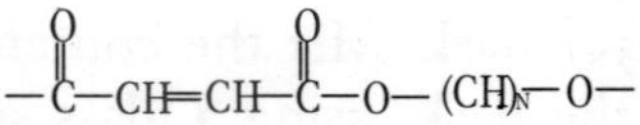 $-C(=O)-CH=CH-C(=O)-O-(CH_2)_N-O-$	crosslinked with styrene and peroxide: fiberglass, boat resin
Cellulose Acetate	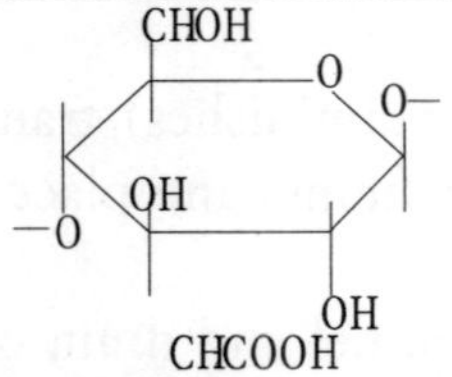 CHOH, O, O–, OH, –O, OH, CHCOOH (cellulose ring)	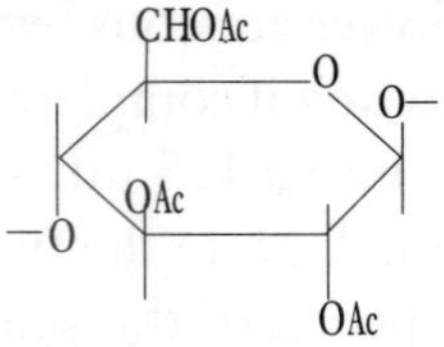 CHOAc, O, O–, OAc, –O, OAc (acetylated ring)	photographic film
Silicones	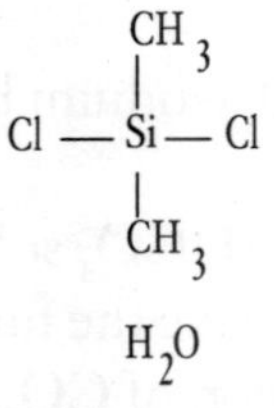 $Cl-Si(CH_3)_2-Cl$ H_2O	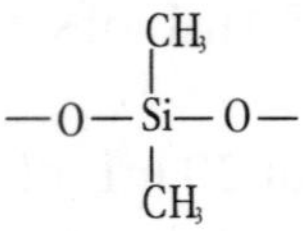 $-O-Si(CH_3)_2-O-$	water-repellant coatings, temperature resistant fluids, rubbers (CH_3SiCl_3 crosslinks in water)
Polyurethanes	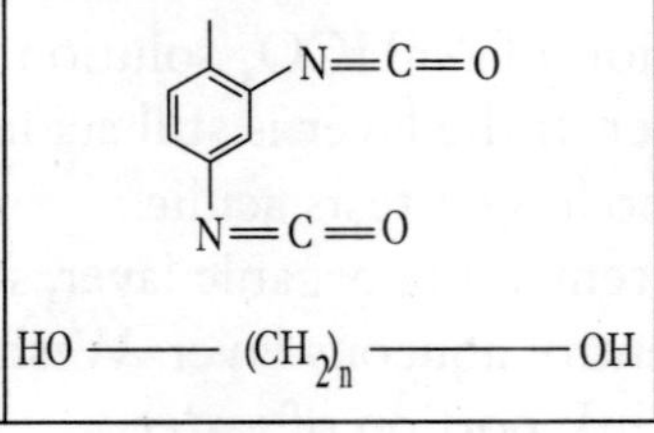 $CH_3-C_6H_3(-N=C=O)_2$ $HO-(CH_2)_n-OH$	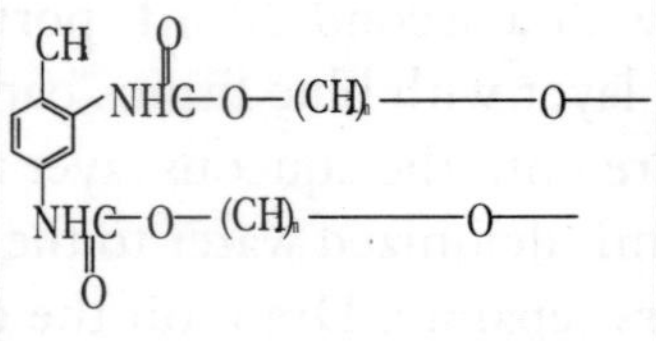 $CH_3-C_6H_3(-NHC(=O)-O-(CH_2)_n-O-)_2$	rigid and flexible foams, fibers

Experimental Procedure

Your TA will assign you 3 out of the 5 sections to perform.

Caution Wear Eye Protection

Part A Cross-Linked Condensation Polymer

Caution: Steps 1-4 result in corrosive HCl fumes being given off, and should be carried out in a hood:

1. Fill a 25 mL buret with deionized water.
   Caution: The $(CH_3)_2$ $SiCl_2$ solution used in step 2 is corrosive. Avoid contact with skin or clothing.
2. Transfer about 60 mL of $(CH_3)_2$ $SiCl_2$ (dimethyldichlorosilane) petroleum-ether solution to a clean and dry 250 mL Erlenmeyer flask.
   Caution: A too rapid addition of water during step 3 results in a too rapid reaction, the contents spurting out of the flask.
3. Add dropwise about 40 mL deionized water from the buret to the Erlenmeyer flask. Mix the contents of the flask during this addition by swirling the flask, using a glass stirring rod, or a magnetic stirrer. The hydrogen chloride will bubble out of solution at the interface of the upper organic and lower aqueous levels.
4. When the reaction if complete (bubbling diminishes), transfer the contents of the beaker into a 125 mL separatory funnel and place a stopper in the top of the funnel. Note the two layers.
5. Open the stopcock of the separatory funnel and drain off the unwanted lower aqueous layer. Close the stopcock immediately after a few drops of the upper organic layer have drained out.
6. Transfer about 50 mL of saturated $NaHCO_3$ (sodium hydrogen carbonate) solution.
7. Slowly add about 20 mL of the saturated $NaHCO_3$ solution to the organic Layer remaining in the separatory funnel. Shake the funnel, vent the funnel to release excess pressure due to the formation of CO_2 and allow the layers to separate. Drain off the aqueous layer into a 100 mL beaker.
8. Repeat with a second 20 mL portion of $NaHCO_3$ solution and test the aqueous layer with blue litmus paper. If the layer is still acidic, repeat the procedure until the aqueous layer no longer tests acidic.
9. Add 10 mL deionized water to the remaining organic layer, shake, and let the layers separate. Drain off the entire aqueous layer. Wash the organic layer a second time with a fresh 10 mL portion of water.
10. Transfer the remaining organic layer to a clean and dry 125 mL Erlenmeyer flask.

11. Add 1 to 2 g anhydrous Na_2SO_4 (sodium sulfate), shake the contents, cover, and allow the flask to stand for about 30 minutes.
12. Decant the solution into a 15 mL beaker, being careful to leave the hydrated Na_2SO_4 behind.
    Caution: The petroleum ether evaporated in step 13 is a volatile and flammable liquid. Do not use an open flame to effect evaporation.
13. Place the 150 mL beaker into a 400 mL beaker half-filled with hot water obtained from the hot water-water tap, and evaporate the petroleum ether. You may have to add additional hot water.
14. Add about 1.5 b B_2O_3 (boron trioxide) and stir well. Note and record the appearance.
15. Use a spatula to remove the sticky material from the beaker, and then mold the substance into a ball. See if it bounces.
16. Place the ball in a labeled 50 mL beaker, and then heat it in a drying oven (200 °C) for about 2 hours.

Part B: Glyptal™ Resin (linear condensation copolymer)

A small aluminum dish is used to mold the polymer. A coin may be placed on the bottom of the mold if you wish to make a souvenir of this experiment.

1. In a large test tube, mix 4.0 mL glycerol, 0.5 g sodium acetate, and 10.0 g phthalic anhydride.
   Caution: Excessive heating during step 2 can cause the hot contents to spurt out. If the hot liquid contacts the skin, severe burns result.
2. Carefully heat the mixture with a low flame, starting at the top of the contents and moving down toward the bottom as the mixture melts.
3. Continue heating until the melt appears to boil, and then continue heating for 5 minutes. Sufficient heating is required to produce a nonsticky product but excessive heating turns the product into a brittle amber material.
4. If desired, place a clean and dry coin into the aluminum dish used for a mold.
   Caution: Avoid skin contact with the hot, sticky liquid transferred in step 5, as contact causes severe burns.
5. Carefully pour the hot liquid into the mold.
6. Allow the material to cool until the end of the laboratory period. (Meanwhile, proceed with Part C.)
7. After the material has cooled, peel the mold from the cooled polymer and describe its appearance.

Part C: Polymethylmethacrylate (addition polymer)

1. Half-fill a 400 mL beaker with deionized water and set it aside.

2. Transfer 5.0 mL water into a clean 6-inch (15 cm) test tube, and make a 5.0 mL calibration mark on the tube with a piece of tape or a glass-marking pencil.
3. Drain and thoroughly dry the test tube.
4. Carefully fill the dry test tube up to the 5.0 mL calibration mark with methylmethacrylate.
   Caution: Although the benzoyl peroxide in step 5 is used dermatologically for the treatment of acne, it can explode when heated. Do not exceed the stated amount of benzoyl peroxide. Wear eye protection!
5. Add about 20 mg benzoyl peroxide to the test tube using a small spatula. This amount is approximately equal in volume to that of a pea.
6. Use a ring stand and clamp the test tube in a boiling water bath
7. Stir the contents of the test tube with a clean, dry glass rod until the benzoyl peroxide is dissolved.
   Caution: Avoid breathing the fumes generated in the test tube during step 8. The monomeric methyl methacrylate is toxic.
8. Continue to heat the test tube in the water bath; do not stir until the contents thicken and stop foaming.
9. Remove the test tube from the water bath and slowly pour the thick, warm liquid into the water in the 400 mL beaker.
10. Examine the film that forms on the surface of the water, and describe its appearance.

Part D: Parchment Paper (cross-linked condensation polymer)

Caution: The concentrated sulfuric acid used in step 1 is extremely corrosive. If spilled on skin or clothing, wash the area immediately with large amounts of water:

1. Carefully pour about 30 mL 12 M H_2SO_4 (sulfuric acid) into a 400 mL beaker.
2. Fill a 250 mL beaker with deionized water.
3. Pour about 20 mL 6 M NH_3 into a second 250 mL beaker.
4. Using tongs, immerse a piece of filter paper (cellulose) in the sulfuric acid solution for 15 to 20 second.
5. Remove the filter paper from the sulfuric acid and dip it into the beaker of deionized water to rinse off the excess acid.
6. Dip the paper into the beaker containing the 6 M NH_3.
7. Rinse the filter paper under running water and set it on a watch glass to dry completely.
8. When dry, compare the texture, appearance, and strength of the treated

filter paper with that of an untreated piece. Write on both pieces with a ballpoint pen and note any differences.

Part E: Cross-linked Polyvinyl Alcohol

Caution: The poly(vinyl alcohol) is a fine dust. Avoid inhaling it.

1. Measure 50 mL of poly(vinyl alcohol) solution into a paper cup or small beaker and observe its properties. Vinyl alcohol does not exist. Poly(vinyl alcohol) is prepared by first forming poly(vinyl acetate) from vinyl acetate following by hydrolysis to the alcohol.
2. Measure 7-8 mL of sodium tetraborate solution into another cup or beaker and observe its properties.
3. Pour sodium tetraborate into the poly(vinyl alcohol) solution while stirring vigorously with a wooden stick. The borate forms a complex structure called tetraborate, $B_4O_5(OH)_4^{2-}$, that links the poly(vinyl alcohol) polymer strands together by hydrogen bonds.
4. Examine the properties of the cross-linked polymer. See how far the polymer will flow from your hand. Is the flowing endothermic or exothermic?

Macromolecules and Molecular Diversity

Without water and small organic molecules, life as we know it would not be possible. The structure of these molecules is intimately related to their function. A continuing theme throughout much of the biological world is this relationship between form and function. When these small organic molecules are joined together, "giant" molecules are produced. These giant molecules are known as macromolecules.

Polymers

Macromolecules are polymers. Polymers are large molecules of many similar "units" linked together. These individual units are called monomers.

Form and Function

The variation in the form of macromolecules is largely responsible for molecular diversity. Much of the variation that occurs both within an organism and among organisms can ultimately be traced to differences in macromolecules. Macromolecules can vary from cell to cell in the same organism as well as from one species to the next.

Generally speaking, all macromolecules are produced from a small set of about 50 monomers. Different macromolecules vary because of the arrangement of these monomers. By varying the sequence, an incredibly large variety of macromolecules can be produced.

While polymers are responsible for the molecular "uniqueness" of an organism, the common monomers mentioned above are nearly universal.

Assembling and Disassembling Polymers

While there is variation among the types of polymers found in different organisms, the chemical mechanisms for assembling and disassembling them are largely the same across organisms. Monomers are generally linked together through a process called dehydration synthesis while polymers are disassembled through a process called hydrolysis.

8.2 Biological Macromolecules

Living organisms are not in equilibrium with their surroundings:

- ❖ Chemical reactions run their courses until there is no further change; this is called equilibrium.
- ❖ In biology: equilibrium = death.
- ❖ Continuous input of energy required to keep processes from going to equilibrium.
- ❖ To counteract the trend toward equilibrium cells burn fuel molecules (sugar, fat, amino acids) and use the released energy to make ATP.
- ❖ Cells use energy for many purposes, including building macromolecules (giant molecules) from small building blocks.

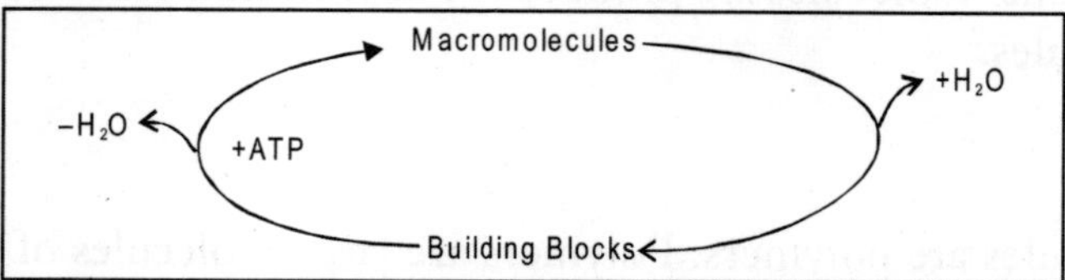

- ❖ Without energy input (from ATP) cells break down into small building block molecules and die (equilibrium state).
- ❖ Each time a bond is formed to make a macromolecule water is removed (dehydration).
- ❖ When macromolecules break down spontaneously they take up water (hydrolysis).
  - This is what happens in digestion.
- ❖ Global picture of energy relationships:
  - Cells exist in a stable relationship with the environment called a steady-state:

- *Example of steady-state 1*: Production of fresh water from sea water using energy from sun (non-living steady-state).
- *Example of steady-state 2*: Production of macromolecules from small molecules using ATP energy.
- *Example of steady-state 3*: Photosynthesis- production of sugars and other carbon compounds from CO2 and water, using light energy (living steady-state).

- In these steady states the system is pushed away from equilibrium by the continuous inflow of energy from outside
- The universe as a whole still running down (that's where the outside energy comes from); we become more complex at the expense of the universe:
    - More than 99 per cent of the energy for life comes from the sun via photosynthesis.

❖ What will happen to a cell with a low supply of energy? Discuss your answer.

AdenosineTriphosphate (ATP) is the Energy Currency of the Cell

❖ Animal cells cannot directly use most forms of energy:

- Most cellular processes require energy stored in the bonds of a molecule, adenosine triphosphate (ATP).
- ATP is referred to as the energy currency of the cell.

This is the structure of ATP. It is a nucleotide, formed from: (a) the base adenine (the structure with 2 rings), (b) the 5 carbon sugar deoxyribose (one ring) (c) 3 phosphates. Energy is stored in the bonds between the phosphates and is released when the bonds are broken.

- Cells are designed to use chemical energy rather than heat energy
  - The cell is not a heat engine- it runs at a constant low temperature
- Energy storage:
  - Body cannot store much ATP (ATP interferes with many chemical reactions)
  - Most of the body's energy is stored in fats (triglycerides: see below) and polysaccharides;
  - ATP is generated from the fats & polysaccharides as needed (this is the purpose of glycolysis & the Krebs cycle):
    - Cell produces energy by oxidizing ("burning") the stored energy rich chemicals,
    - Some of the energy goes off as heat, but most is trapped in the phosphate bonds of ATP.
  - Some electrical energy is also stored as ion gradients.

Cells Contain 4 Major Types of Giant Molecules (Macromolecules or Polymers)

- Proteins
- Polysaccharides (complex sugars)
- Nucleic Acids
- Some Lipids (fats).

Biological Polymers are Made from About 60 Small Building Blocks

- Cells contain thousands of types of giant molecules (macromolecules, polymers).
- Most of the polymers are made from about 60 types of small building block molecules.
- Hooking the building blocks together to make polymers requires energy from ATP:

Building Blocks	*Number of Kinds*	*Polymers*
Amino Acids	20	Proteins
Fatty Acids	10	Storage Lipids/Membranes
Sugars & Relatives	10	Polysaccharides/Nucleic Acids
Nucleotides	5	Nucleic Acids
Others	15	All of Above

- Basic amino acid structure: amino (NH2) and carboxyl (COOH) groups.
- The 20 types have different side groups (designated as R).

❖ In proteins the amino acids are linked head to tail by peptide bonds.

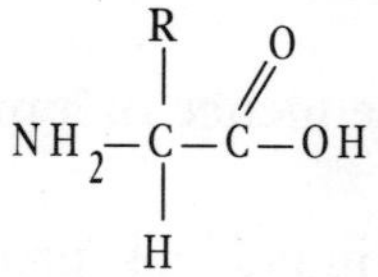

$$NH_2-\underset{\substack{|\\ H}}{\overset{\substack{R\\ |}}{C}}-\overset{\substack{O\\ //}}{C}-OH$$

Basic Amino Acid Structure

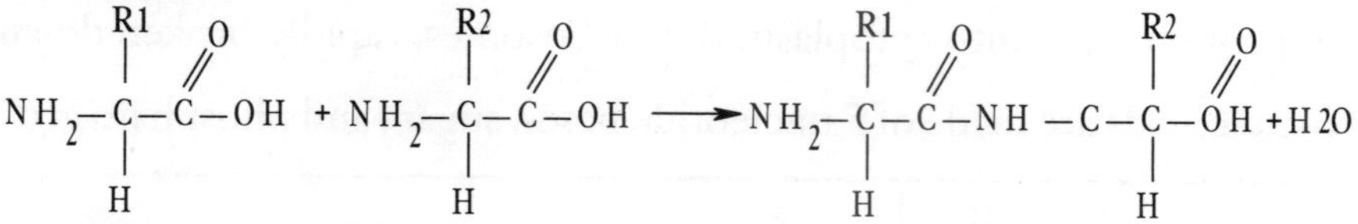

Note the amino (NH2), carboxyl (COOH) and hydrogen groups (H) which are found on all amino acids. The differences in amino acids are in the 20 different types of R groups.

$$NH_2-\underset{\substack{|\\ H}}{\overset{\substack{R1\\ |}}{C}}-\overset{\substack{O\\ //}}{C}-OH + NH_2-\underset{\substack{|\\ H}}{\overset{\substack{R2\\ |}}{C}}-\overset{\substack{O\\ //}}{C}-OH \longrightarrow NH_2-\underset{\substack{|\\ H}}{\overset{\substack{R1\\ |}}{C}}-\overset{\substack{O\\ //}}{C}-NH-\underset{\substack{|\\ H}}{\overset{\substack{R2\\ |}}{C}}-\overset{\substack{O\\ //}}{C}-OH + H2O$$

Formation of a Peptide Bond

Two amino acids can combine, with the carboxyl of one attaching to the amino group of the other. Water is removed. This process also requires energy in the form of ATP.

❖ Proteins differ from one another in 2 ways:
  - *Size*: number of amino acids:
    - ◆ Typical protein has 200 to 1000 amino acids.
  - Sequence of amino acids:
    - ◆ Some amino acids are charged or have a polar structure: hydrophilic (like water).
    - ◆ Some amino acids are neutral and nonpolar: hydrophobic (hate water, like to be in lipid).
    - ◆ Hydrophobic proteins are found in membranes.
  - Hydrophobicity can be tested by oil/water partition test.

❖ Why do you suppose hydrophobic proteins tend to be found in cell membranes? Discuss your answer.

Proteins Do Most of the Work of the Cell

❖ Every cell has thousands of different types of proteins, each specialized to do a certain job.

❖ *Some proteins are structural*: Control shape of cells and bind cells together *Example*: collagen- binds all of the cells of the body together.

❖ Chemical reactions of the cell are controlled by protein enzymes.

❖ Protein pumps move things across the cell membrane.

❖ Proteins give mobility:
  - Muscles
  - Flagella & cilia.
  - "Molecular motors"

❖ Defend the body against foreign invaders: antibodies.

❖ *Receptors*: required for signaling in endocrine and nervous systems.

Nucleic Acids are the Molecules of Heredity

❖ Two major types: DNA & RNA:

- Both types have code which specifies the sequence of amino acids in proteins
- DNA = archival copy of genetic code, kept in nucleus, protected
- RNA = working copy of code, used to translate a specific gene into a protein, goes into cytoplasm & to ribosomes, rapidly broken down

❖ Nucleic acids are made of 5 nucleotide bases, sugars and phosphate groups:

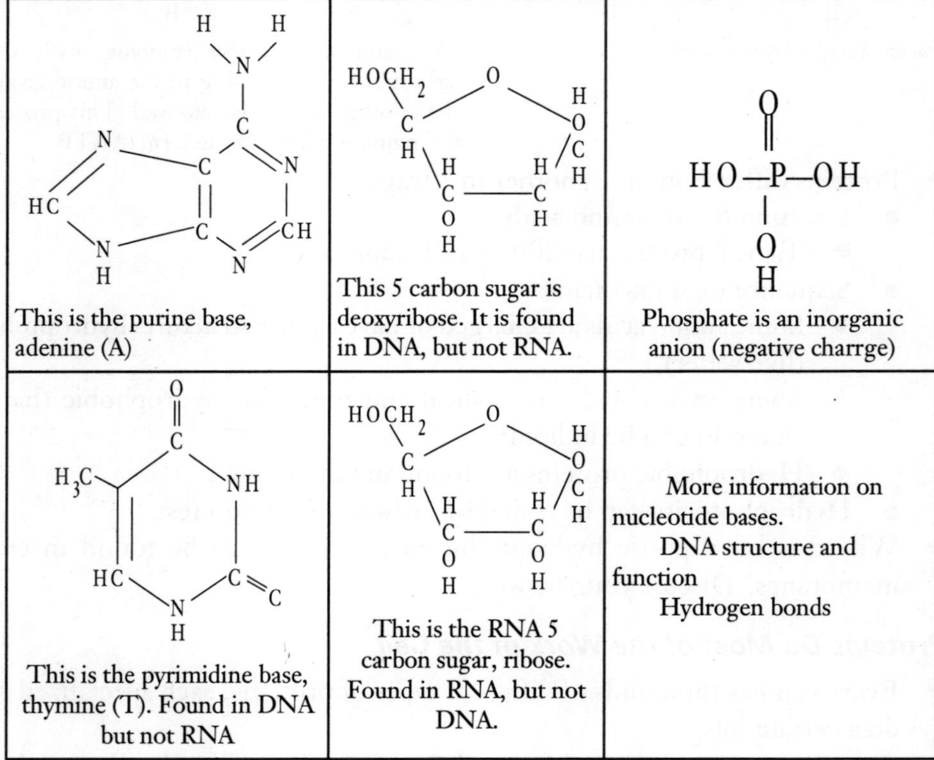

This is the purine base, adenine (A)

This 5 carbon sugar is deoxyribose. It is found in DNA, but not RNA.

Phosphate is an inorganic anion (negative charrge)

This is the pyrimidine base, thymine (T). Found in DNA but not RNA

This is the RNA 5 carbon sugar, ribose. Found in RNA, but not DNA.

More information on nucleotide bases.
DNA structure and function
Hydrogen bonds

❖ Summary of the differences between DNA and RNA:

	Strands	*Code*	*Nucleotide Bases*	*5 Carbon Sugar (a pentose sugar)*	*Phosphate*
DNA	2	Archive	A, C, G, T	Deoxyribose	Yes
RNA	1	Working copy	A, C, G, U	Ribose	Yes

❖ The bases make up the genetic code; the phosphate and sugar make up the backbone.

❖ RNA is a molecule with a single strand.

❖ DNA is a double strand (a double helix) held together by hydrogen bonds between the bases.

- A = T; C= G because:
  - ◆ A must always hydrogen bond to T.
  - ◆ C must always hydrogen bond to G.

Lipids Form Cell Membranes and Energy Storage Depots

❖ Lipids are a group of unrelated molecules with these properties:

- Insoluble in water.
- High oil/water partition coefficients.
- Oil/water partition illustrated with colored dyes:

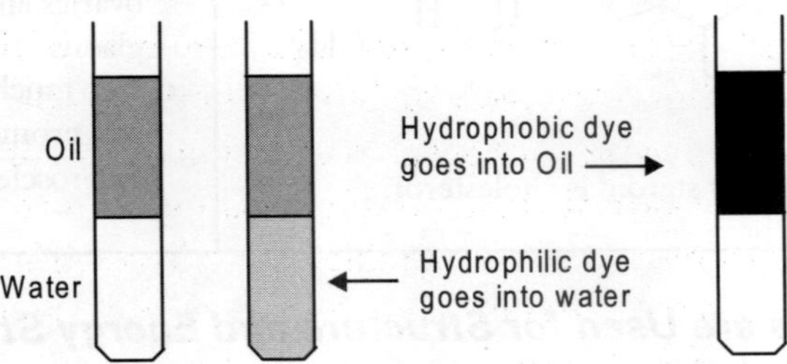

- TThe oil/water partition coefficient = amt. of substance in oil/amt of substance in water.
  - ◆ Hydrophobic substances have high oil/water partition coefficients
  - ◆ In the body, hydrophobic molecules tend to accumulate in fat tissue
  - ◆ Hydrophobic drugs cross cell membranes faster than water soluble drugs .

❖ *Triglycerides*: storage fat:

- Composed of 3 fatty acids & glycerol:

```
     H
     |    O
     |     \\   H H H H H H H H H H H H H H
H—C—O—C— C—C—C—C—C—C—C—C—C—C—C—C—C—C—H
     |          H H H H H H H H H H H H H H
     |
     |    O
     |     \\   H H H H H H H H H H H H H H
H—C—O—C— C—C—C—C—C—C—C—C—C—C—C—C—C—C—H
     |          H H H H H H H H H H H H H H
     |
     |    O
     |     \\   H H H H H H H H H H H H H H
H—C—O—C— C—C—C—C—C—C—C—C—C—C—C—C—C—C—H
     |          H H H H H H H H H H H H H H
     H
```

This is an example of a triglyceride. To the right is a 3 carbon glycerol molecule. Attached to it are 3 fatty acids with long chains. Triglycerides are the main storage form of energy in the body. You can store 9 Calories per gram as triglycerides.

- Phospholipids: form cell membranes:
  - Composed of 2 fatty acids, glycerol, phosphate and polar groups
  - Like triglycerides, but one of the 3 fatty acids is replaced by polar groups.
- *Steroids*: have 4 rings- cholesterol, some hormones, found in membranes:
  - Sex hormones, cortisone and aldosterone are steroids:

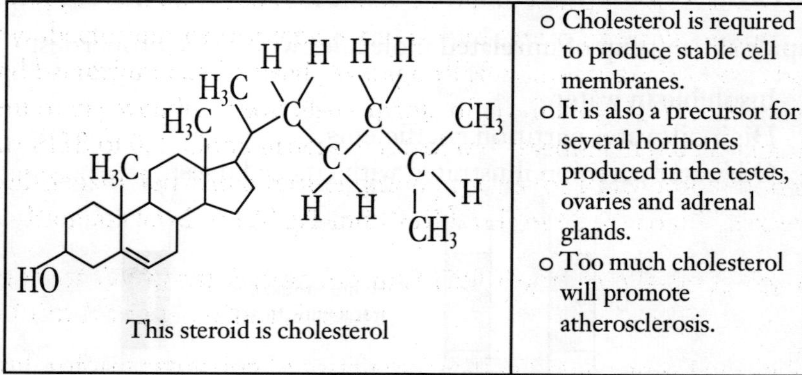

This steroid is cholesterol

- o Cholesterol is required to produce stable cell membranes.
- o It is also a precursor for several hormones produced in the testes, ovaries and adrenal glands.
- o Too much cholesterol will promote atherosclerosis.

Polysaccharides are Used for Structure and Energy Storage

- Polysaccharides are polymers made of sugar molecules linked together:
  - Glycogen: energy storage in animals (4 Calories/gram): made of glucose
    Starch: energy storage in plants: also made of glucose
- Glucose is most important sugar (6 carbons = a hexose sugar):

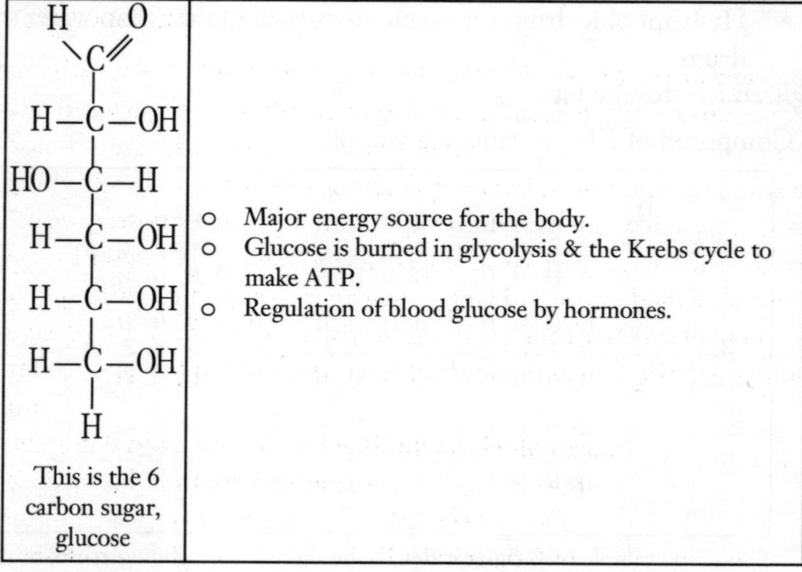

This is the 6 carbon sugar, glucose

- o Major energy source for the body.
- o Glucose is burned in glycolysis & the Krebs cycle to make ATP.
- o Regulation of blood glucose by hormones.

  - Major source of cell energy.

- The brain is very sensitive to the blood glucose concentration because it cannot use fats for an energy supply.

❖ Some sugars are attached to membrane proteins (example: ABO blood groups).

Summary of Biological Macromolecules

Macromolecule	*Building Blocks*	*Functions*
Polysaccharides	Sugars	Energy storage (4 Cal/gm) Structure (cell walls, exoskeletons)
Lipids: triglycerides	Fatty acids, glycerol	Energy storage (9 Cal/gm)
Lipids: phospholipids	Fatty acids, glycerol, phosphate, polar groups	Cell membranes
Proteins	Amino acids: 20 types	Cell structure Enzymes Molecular motors (muscle, etc) Membrane pumps & channels Hormones & receptors Immune system: antibodies
Nucleic Acids: DNA (forms a double helix)	4 Bases: A, C, G, T Deoxyribose sugar, phosphate	Storage of hereditary information (genetic code)
Nucleic Acids: RNA 3 types: m-RNA t-RNA r-RNA (usually a single strand)	4 Bases: A, C, G, U Ribose sugar, phosphate	Protein synthesis: • m-RNA: working copy of genetic code for a gene (transcription) • t-RNA & r-RNA: translation of the code

8.3 Biochemistry

The Bio-Macromolecules we will look at are polymers (poly = many; mer = subunit). They are composed of a subunit bonded to a subunit bonded to a subunit ... on and on and on. The number of subunits in these macromolecules ranges from ten or twenty up to billions for the Nucleic Acids.

Almost all of the molecules that make up your body are polymers, chains of subunits. There are four fundamental types of Bio-macromolecules. Each type of macromolecule is a polymer composed of a different type of subunit. Identifying the type of macromolecule by their subunit composition is a technique you want to become familiar with. Once we understand the chemocal properties of the subunits we are well on our way to understanding the chemical properties of Bio-molecules.

The subunits of the macromolecules we will examine are covalently-bonded. The covalent bonds between the subunits are always formed by a type of reaction called *Dehydration Synthesis* (making something while losing water).

ਣ ਕੀ ਖਦਗ਼ ਢਬਅ ਛ ਖਅਵੈ ਕਤਜ਼ਬ

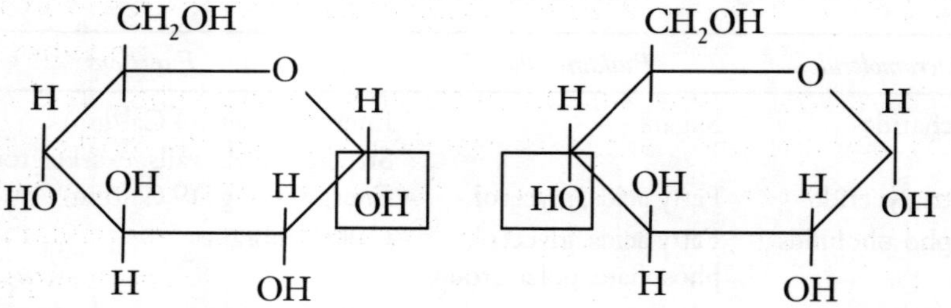

ਰਿਲ਼ . ਲਰਅਦ ਮੀ ਜ਼ਕ
;ਰਤਜਅਪ. ਮ. ਢਕਗਠ ਰ;ਕਫਚ;ਕ

During dehydration synthesis, a water molecule is lost, electrons are rearranged, and a new bond is formed between the subunits.This same tyoe of reaction is used to join the subunits of each of types of large molecules (macromolecules) that make organisms what they are. The structure of each type of macromolecules is what determines how the molecules function in our bodies. Remember to concentrate on how structure and function are related as you examine the composition of these molecules in the following pages.

The first group that we examine is the *Carbohydrates.* Later we will look at:

❖ Proteins,
❖ Lipids, and
❖ and Nucleic Acids.

8.4 Biological Interfaces

We are interested in the study of biological interfaces and their applications in medicine. This involves:

❖ Interfacial phenomena in physiological systems.
❖ Evaluation of the interactions at biological interfaces like the saliva ?mucosa interface in the oral cavity.
❖ Evaluation of the interactions at biological interfaces like the saliva ?mucosa interface in the oral cavity.
❖ Use of the Langmuir and Langmuir Blodgett models for studying drug penetration through cell membranes.

This research leads to a better understanding of physiological and diseased states at a monomolecular level. Also, dynamic surface tension and adsorption kinetics of biological macromolecules and bio-fluids are used for diagnostic and therapeutic purposes.

Drug Penetration Studies

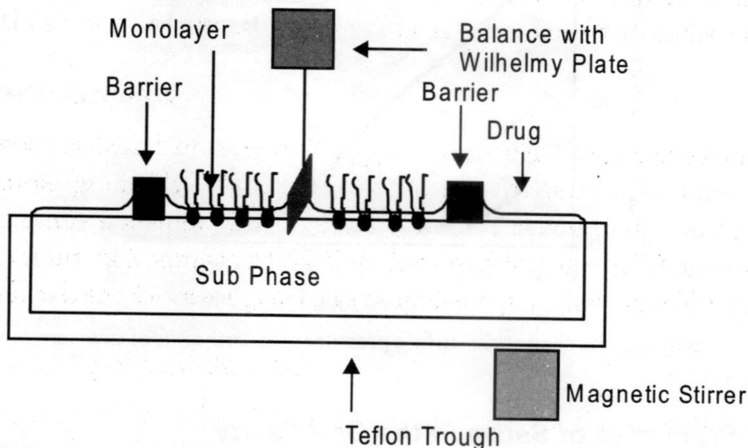

Fig. 8.1: Figure depicts the schematic of drug penetration studies using Langmuir monolayers.

The penetration profiles of drugs as well as drug delivery carriers can be evaluated by monitoring the kinetics of change in surface pressure due to insertion of the drug/carrier in a Langmuir monolayer. The monolayer composition is chosen similar to that of a model cell membrane and is compressed to the required surface pressure prior to injection of the drug in the subphase. Important pharmacotoxicological and thermodynamic parameters regarding the interaction between the drug and the monolayer molecules can be derived from such studies.

8.5 Interfacial Properties of Cell Membranes in Cervical Cancer

The underlying hypothesis for the study is that the oncogenes of hormone responsive tumors cause an induction of certain phosphorylation mechanisms that alter the phospholipid profile of the cancerous tissue as opposed to the normal. This in turn is detectable as subtle changes in the surface activity and state of the cancerous tissue. We have established interesting evidence of the altered phase states and surface activity profiles of cancerous and normal tissues. The role of individual tissue components tissues on tensiometric profiles has also been analyzed. We have found that it is possible to fingerprint cancerous and normal tissues using specific tensiometric parameters. Our results pave the way for developing strategies towards modulating interfacial properties for therapeutic benefit. In short, ease of technique, precision and low standard deviations within a group make tensiometric profiles promising as a possible biophysical marker in cancer research.

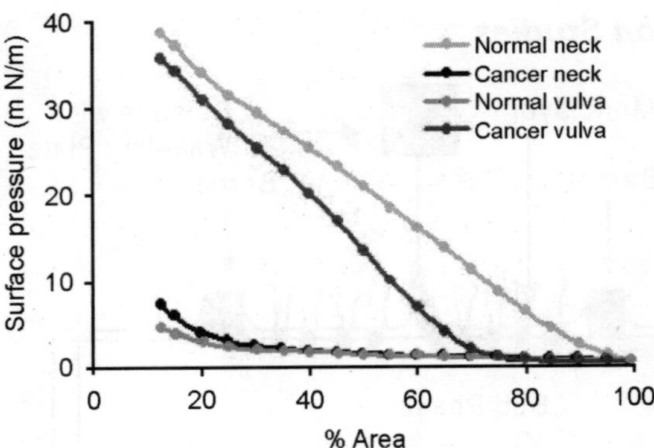

Fig. 8.2: Figure depicts the changes in surface pressure area isotherms of canacerous and normal tissues.

Surface Properties of Saliva in the Oral Cavity

The formation of an adequate film of saliva in the oral cavity is essential to the many functions that it needs to fulfill like preventing dehydration of the mucosa, lubrication of the surfaces and help in speech and taste sensation. This is determined by the surface tension, spreading and adsorption of salivary components on the oral mucosa. In various cases of salivary gland dysfunction and post-radiotherapy, saliva substitutes are required for therapeutic benefits. Substitutes must be evaluated in terms of specific interfacial and viscoelastic criteria based on normal saliva. We have established the desired biophysical criteria for artificial saliva based on the properties of natural saliva. We are evaluating existing substitutes with regard to the required interfacial criteria. Research is ongoing for the development of better artificial saliva substitutes based on biophysical properties.

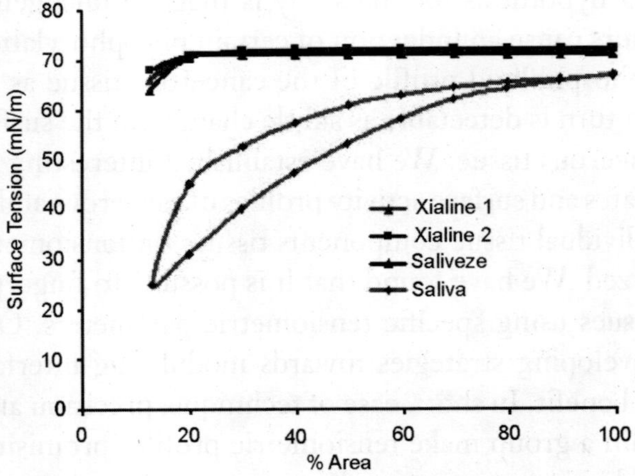

Fig. 8.3: Figure depicts the surface tension of saliva and saliva substitutes.

Cell Adhesion On Biomaterials

The surface parameters that determine the attachment of certain cells to biomaterial surfaces is of interest to us. This in turn predicts the compatibility of the substance. The wettability and the work of adhesion using different surfaces as well as fluids of increasing physiological complexity are being used to predict cell adhesion. The ability to repel platelets and erythrocytes is specifically being evaluated.

8.6 MSc(Eng) Polymers and Polymer Composites Science and Engineering or PhD with Integrated Studies

The MSc(Eng) in "Polymers" was established more than 20 years ago and has successfully educated graduates from across the world. Active research with polymeric materials has informed the curriculum. Recently, the University of Sheffield decided to promote the study of Polymer Science and Engineering with the appointment of a Polymer Materials Scientist and a Polymer Chemist to Chairs in the Department of Chemistry and a Polymer Physicist to a Chair in the Department of Physics. As a result The Sheffield Polymer Centre was formed in 2000. Furthermore the Sheffield Polymer Centre is now part of the IRC in Polymers which was originally formed in the Universities of Leeds, Bradford and Durham.

What does the Course Provide?

Common Core Modules include:There are two options in this interdisciplinary course. MSc(Eng) in Polymers and Polymer Composites Science and Engineering and MSc in Polymers for Advanced Technologies:

- ❖ Fundamental Polymer Chemistry;
- ❖ The Physics of Polymers;
- ❖ Biopolymers and Biomaterials;
- ❖ Polymer characterisation and analysis; and
- ❖ Polymer Material Science and Engineering.

Additional Core Modules for MSc in Polymers and Polymer Composites Science and Engineering:

- ❖ Polymer Fibre Composite Materials;
- ❖ Design and Manufacture of Composites or
- ❖ Biomedical Applications of Polymers.

Additional Core Modules for MSc in Polymers for Advanced Technologies (Chemistry Stream):

- Design and Synthesis of Polymers with controlled structures;
- Smart Polymers and Polymeric Materials.

Additional Core Modules for MSc in Polymers for Advanced Technologies (Physics Stream):

- Macromolecules of Interfaces and Films;
- Electronic and Photonic Molecular Materials and Devices.

The students may register for alternative modules after discussions with the course tutors.

Students are expected to complete a research project in a related area.

Recently Completed Projects Include

- Ultraviolet Initiated Surface Grafting of N-isopropylacrylamide for Biomedical Application.
- Microwave absorption by nano-particulate conducting polymer composites.
- New methods for producing auxetic foams and multifunctional foams.
- The effect of a high temperature stage on the strength of sized glass fibres.
- Curing monitoring of carbon fibre-epoxy composites using resistance-based sensors.
- Plasma polymerisation and characterisation of ethanethiol.
- Patterning Techniques for Plasma Polymers.
- Development of microwave smart window composite.
- Stress transfer in single fibre model composite.
- An investigation into developing a self-monitoring system for resin impregnation during VARTM using the change in resistance of carbon fibres.
- Probing the dense packing of hyperbranched polymers.
- Synthesis of vinyl ether oligomers with ester end groups.
- Electroplating onto synthetic metals.
- Synthesis and characterisation of Thiol-Terminated polystyrene block copolymers.
- Novel acrylonitrile copolymers.

Overview of MSc Courses in "Polymers" within the Sheffield Polymer Centre

The Course is taught by experts in all aspects and is structured with a common core for all students:

- CHM6106 Fundamental Polymer Chemistry
- MAT6102 The Physics of Polymers

- ❖ CHM6108 Biopolymers and Biomaterials
- ❖ CHM6201 Polymer Characterisation and Analysis
- ❖ MAT6103 Polymer Material Science and Engineering
- ❖ MAT6030 Competitive Materials/Essay

MSc(Eng) in Polymer and Polymer Composite Science and Engineering (MAT03)

Additional Core Modules:

- ❖ MAT6430 Polymer Fibre Composite Materials
- ❖ MAT6040 Research Project in the Department of Engineering Materials

Optional Modules:

- ❖ MAT6104 Design and Manufacture of Composites (recommended)
  Alternatively you may choose one of the following:
- ❖ MAT6203 Biomedical Applications of Polymers
- ❖ CHM6204 Smart Polymers and Polymeric Materials
- ❖ PHY6006 Macromolecules at Interfaces and Films
- ❖ PHY600 Electronic & Photonic Molecular Materials and Devices

MSc Polymers for Advanced Technologies (Chemistry Stream)

Additional Core Modules:

- ❖ CHM6202 Design and Synthesis of Polymers Controlled Structure
- ❖ CHM6301 Polymer Research Project in the Department of Chemistry

Optional Modules:

- ❖ CHM6204 Smart Polymers and Polymeric Materials (Recommended)
  Alternatively you may choose one of the following:
- ❖ MAT6203 Biomedical Applications of Polymers
- ❖ MAT6430 Polymer Fibre Composites
- ❖ MAT6104 Design and Manufacture of Composites
- ❖ PHY6006 Macromolecules at Interfaces and Films
- ❖ PHY6007 Electronic & Photonic Molecular Materials and Devices

MSc Polymers for Advanced Technologies (Physics Stream) (CHMT01)

Additional Core Modules:

- ❖ PHY606 Macromolecules at Interfaces and Films

- ❖ PHY6007 Electronic & Photonic Molecular Materials and Devices
- ❖ PHY6009 Research Project in the Department of Physics.

Diploma in Polymers and Polymer Composite Science and Engineering/Diploma in Polymers for Advanced Technologies

The students follow the above course structures but do not complete a research project and submit a dissertation. Students who perform at a standard appropriate to the MSc course, in Semester 1 examinations, can apply to be upgraded.

Why Study Polymers at the University of Sheffield

The Sheffield Polymer Centre has a wide range of facilities across the contributing departments, with specialists in most facets of the subject. The Department of Engineering Materials achieved 5*A rating in the research assessment exercises in 1996 and 2001, while the Departments of Physics and Chemistry were rated 5B in 2001. These demonstrate the strength, depth and quality of the staff.

What Next?

You will acquire skills of direct use to a range of employers and for research PhD´s across the Polymer Centre.

Alternative Options:

PhD with Integrated Studies in Advanced Materials (Polymers)

The PhD with Integrated Studies is offered at the University of Sheffield as part of the New Route PhD initiative—a major development involving over 20 leading UK universities and supported by the Higher Education Funding Council for England and the British Council. Leading to a full PhD thesis, the programme is designed to enhance students' subject knowledge and develop a range of subject-specific and important generic research skills. Responding to requests from our international partners for a PhD programme with a greater taught component, the PhD with Integrated Studies will also provide students with additional skills for the world of work and so enhance their employability on completion of their programme. The PhD with Integrated Studies [Advanced Materials (Polymers)] is a 4 year course.

8.7 Dendritic Macromolecules at Interfaces

The ultimate goal of this project is to understand the interfacial behavior and microstructure of monodendrons in order to immobilize these molecules on an interface. We are going to screen various molecular parameters including a generation number of monodendrons, type of terminal groups, spacer length, nature of anchoring groups, and type of functionalized substrates in a search for stable organized interfacial assemblies. These dendrimer macromolecules could be capable of forming organized and stable supramolecular structures at interface with the focal anchoring group used for chemical binding to functionally modified surfaces. Interfacial ordering and surface behavior at air-liquid, air-solid, and liquid-solid interfaces will be studied by Langmuir technique, a chemical and electrostatic self-assembly in conjunction with comprehensive characterization with ellipsometry, x-ray scattering, UV spectroscopy, and scanning probe microscopy. The project focuses on the nanotechnology for fabrication of nanocomposite films with molecularly tailored surface properties from dendrimers. Dendrimer polymers with a tree-like architecture are promising candidates for functional supramolecular materials with prospective applications in the fields of drug delivery, molecular coatings, nanochemistry, molecular composites, and light harvesting. A critical step in the development of such an application is the understanding of how these macromolecules can be anchored to solid substrates in a controllable manner. Two undergraduate students will be involved in these studies of advanced polymer coatings from functional dendrimer materials.

8.8 Dynamics of Polymer Ultrathin Films

Under the effect of confinement and specific interactions with the interfaces, mechanical and dynamical properties of a polymer may be perturbed notably. These changes are expected to occur to thin films with thickness comparable to the size of the polymer molecules, which is typically of tens of nanometers. We refer to these films as *ultrathin* films. Experimentally, it was observed that the diffusion coefficient of polymers in a thin film decreased precipitously as the film thickness was decreased below about the size of the macromolecules; the glass transition temperature, Tg was also found to be a strong function of the film thickness. We are interested in the dynamical regime below and near the Tg. A number of new techniques were developed to probe the dynamic behaviors of the ultrathin polymer film systems. Experiments were carried out to understand the factors affecting the dynamic behaviors.

I. Dynamics Revealed by Atomic Force Microscopic Adhesion Measurements (AFMAM)

By using AFMAM, we were able to directly probe the ac mechanical response of the polymer thin film over a wide dynamical range near the *Tg*, which at present is only possible with AFM Lateral Force Microscopy. Effort is now being put into extending the present understanding of the technique into studying polymer films of different thicknesses and correlating the results with the measured T_g of the films from other techniques such as X-ray Reflectivity and Ellipsometry.

During measurement, force-distance (FD) curves were acquired with the AFM. Adhesion force, *Fad* is determined from the pull off leg of the force-distance curves as the pull off force required to cause detachment between the AFM tip and the sample surface.

Figure below shows the FD curves obtained as a poly (tert-butyl acrylate) (PtBuA) film (*Tg* ~ 50°C) evolves from the glassy to the rubbery state (top to bottom). Development of curvatures in the FD curves can be attributed to softening in the polymer as the rubbery state is approached, either by increasing the temperature, *T*, or lowering the probe rate, *f*, (i.e. the rate at which the tip probes into the sample and being pulled out). However, what is immediately evident from the data is the time-temperature equivalence of the dynamics revealed by the FD curves, wherein FD curves that are acquired at equivalent time-temperature conditions (solid and dashed lines) are essentially the same.

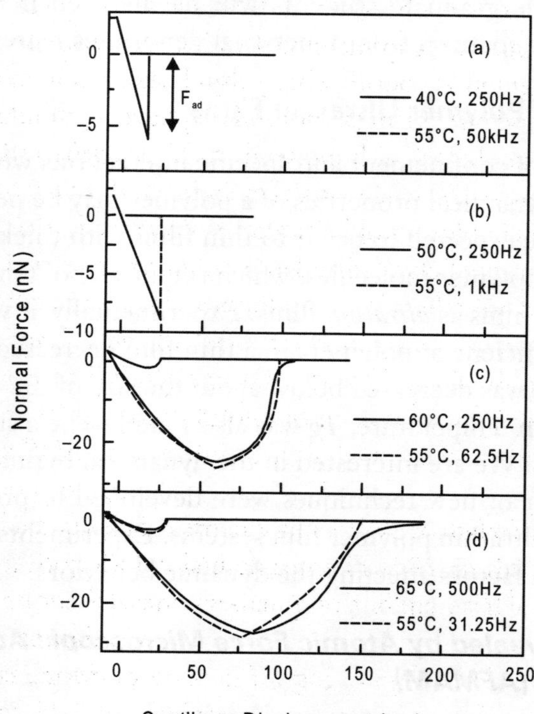

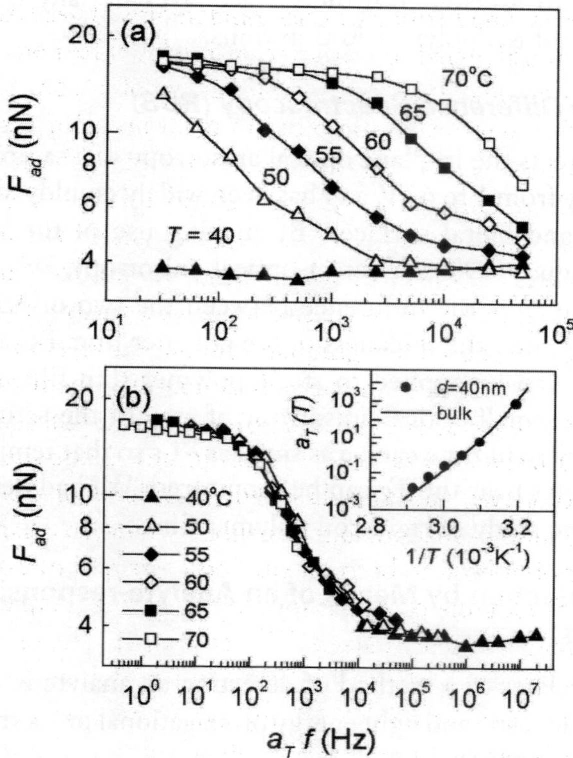

A plot of *Fad* vs. *f* at different *T* near the *Tg* [Fig. above (a)] further confirms the time-temperature equivalence characteristics of the dynamics being probed. Upon rescaling the abscissa of each *Fad* (f) curve by a temperature-dependent shift factor, *aT(T)*, an adhesion master curve is obtained [Fig. above (b)]. The temperature dependence of the shift factors obtained can then be compared to the bulk data to check for any deviation.

Detailed analysis of the mechanical problem reveal that the shape of the adhesion master curve in the high-frequency regime (above the low- frequency plateau region) is proportional to the reciprocal of the time-dependent elastic modulus of the polymer, *E(t)*. The plateau region, however, is where the sample-tip bond fracture occurs cohesively so should be considered separately from the behavior in the high-frequency regime.

II. Effect of a High-mobility Surface Layer at the Polymer/Air Interface

Experimental and theoretical findings suggest that the segmental mobility of polymers with low molecular weight and/or with low surface-energy end groups is enhanced at the polymer-air interface. We have undertaken systematic studies on the glass transition temperature of polymer ultrathin films with different surface mobility to investigate if the surface layer has any influence on the overall dynamics of the polymer in thin films.

III. Reflectance Difference Spectroscopy (RDS)

This method detects the in-plane optical anisotropy of a sample as a function of photon energy from 1 to 6 eV, and has been widely employed in the study of semiconductor and metal surfaces. By making use of the phase-sensitive detection technique, RDS can detect optical anisotropy, ?*n***t* as small as 0.2 Angstrom, where ?*n* is the difference between the two orthogonal in-plane refractive indices and *t* the thickness of the polymer film. Because of the high sensitivity, RDS can be applied to study polymer thin film samples, which typically has very small optical anisotropy, if any. In the setup we currently have, the measurement time can be as small as ~1 s so that temporal relaxation of polymer samples near the *Tg* can be monitored. We had demonstrated the use of RDS in the study of rigid-rod polymer films.

8.9 Analyte Detection by Means of an Analyte-responsive-polymer Technical Field

The invention relates to a method of detecting an analyte in a liquid sample that-utilizes the acoustic and light energy propagational properties of a polymer interfaced with a sensor.

Detection mechanisms capable of detecting chemical and biological substances in the nanomolar range are finding increasing utility in a variety of commercial fields including the food, healthcare, environmental and waste treatment industries. A number of sensitive devices have been developed to meet these detection needs, including those taking advantage of shifts in resonant frequency such as piezoelectric transducers as well as optical sensors which operate on the basis of changes in refractive indices.

The ability of piezoelectric transducers to detect small changes in mass, viscosity and density at their surfaces while immersed in liquids has made them particularly useful as analytical tools where the measurement of very small amounts of material must be made in solution. Design strategies for piezoelectric sensors have primarily involved events that result in mass changes at the surface of the crystal. The most extensively studied transducer in this regard has been the shear mode AT-cut quartz resonator, commonly referred to as the quartz crystal microbalance (QCM), which comprises an AT-cut quartz crystal sandwiched between two metal excitation electrodes. An example of the use of the QCM used to detect biological analytes is seen in EPO Patent Publication No. 0 215 669 (Karube et al.) where the concentration of an analyte in solution was calculated on the basis of the change in resonance frequency (Äf) caused by the weight of an analyte added to a receptor material immobilized on the surface of a piezoelectric device. The invention of Karube demonstrated that the piezoelectric transducer was useful in measuring the concentration of some biological analytes.

Although a useful tool in solution environments, a piezoelectric transducer must have its surface modified in order to give it a measure of specificity for the analyte to be detected, and preparation of these modified piezoelectric devices is procedurally complex and often difficult. In cases where the analytes to be detected are biological in nature, receptor agents (antigens, antibodies or other ligands) must be immobilized on the surface of the crystal. This method has significant inherent limitations; however, as the receptor agents can inactivate during the immobilization process or separate from the surface of the crystal after immobilization.

To facilitate receptor modification of the QCM, polymer coated quartz oscillators have been developed that that allow for more efficient and specific binding of receptor agents. Muller-Schulte (DE 3733985 Al) describes the coating of a piezoelectric oscillator with a water-insoluble polymer that facilitates the adsorbtion of various biomolecules. These immobilized biomolecules can then bind antigens or ligands and the resulting increase in mass of the oscillator quartz translates into changes in resonant frequency which can be used to quantitate the analyte.

These examples of the use of the piezoelectric sensors are useful but limited in that they all rely on a mass change on the sensor surface to effect a change in resonant frequency. Sensor design based solely on mass changes can be limiting if the analytes have low molecular mass. For example, the mass increase associated with the binding of a protein to an active surface of a shear mode AT-cut crystal generally will not be sufficient for a practical frequency response. Other considerations include the rigidity of the bound analyte; for example, Newtonian films that bind to the surface of a QCM give frequency shifts that are much smaller than predicted by the Sauerbrey equation (Sauerbrey, Phys. (1959) 155:206).

Additionally, piezoelectric oscillator techniques have not been well-suited to continuous real-time (kinetic) measurements of the biological activity of organisms and the usefulness of receptor-modified piezoelectric methods has been limited by the specificity of known reagents, a fact that requires repeated testing to assess the presence of different organism types. To overcome these limitations, polymer films have been designed to react with specific reagents, ions or metabolites in an attempt to increase the specificity of the detection method. For example, U.S. Patent 4,735,887 (Foss *et al.*) discloses that propyleneimine will react with polyampholytes via a ring-opening mechanism to form primary amines. The primary amines can react with tanning developers and with ordinary aldehyde crosslinking agents to form crosslinked networks. Ebersole et al. (Int'l. Pub. No. WO 91/01381) describe the use of polyampholytes in solution which, upon contact with the metabolic products of an organism, deposit upon a piezoelectric device to produce a change in resonance frequency that is correlated to the concentration or rate of change in the presence of a metabolic product. Here the polyampholyte is not immobilized on the surface of the sensor before exposure to an analyte suspected of being present in a sample. Tanaka et al. (U.S. Patent 4,732,930) have demonstrated that certain ionic gels formed by the polymerization of isopropylacrylamide in the presence of a metal ion containing monomer, crosslinking agent, and a suitable liquid medium are capable of drastic volume changes in response to changes in solvent composition, temperature and pH or ion composition. Tanaka et al., however, make no attempt to link the nature of these gels to a measuring device utilizing piezoelectric oscillators or detection of changes in refractive indices to achieve analyte sensitivity.

Attempts to integrate the dynamic physical dimensional properties of polymer films into analyte detection systems have been few and poorly developed. Various physical dimensional changes have been reported for crosslinked polymers responding to changes in partial pressures of gases such as oxygen (Irani et al., Flammability and Sensitivity of Materials in Oxygen-Enriched Atmospheres: Third Volume, ASTM STP 986, D. W. Schroll, Ed., American Society for Testing and Material, Philadelphia, (1988) pp 346-358) as well as to changes in humidity and salt concentrations. JP 63-206653 describes a salt concentration sensor characterized by the fact that an organic polymer gel layer or mass is retained on a support whigh gives rise to a change of phase or a change in volume depending on the salt concentration. JP 2-212744 describes a semiconductor humidity sensor which detects atmospheric humidity on the basis of the expansion or contraction of a moisture-sensitive polymer in contact with the surface of a piezoelectric cell. Changes in the piezoresistance of the cell are translated into percent atmospheric humidity.

Although gels and crosslinked polymers are known to be responsive to various biological and ionic analytes as shown in the above art, in all cases the preparations depend on a common physical principle of polymer swelling in response to salt or humidity changes.

Furthermore, there is nothing inherent in the polymer gel that makes it specifically responsive to water or a particular ion and all gels must be in a pre-swollen state in order to function. Optical sensors for the detection of analytes generally rely on small changes in the indices of refraction in response to the presence of an analyte. Commonly used optical sensors include planar waveguides, optical fibers and diffraction gratings. In general, optical sensors of this sort suffer from many of the same disadvantages as piezoelectric oscillators such as lack of specificity for an analyte, difficulty in surface preparation as well as difficulty in measuring continuous real-time (kinetic) changes in biological activity. To date, these difficulties in the field of optical sensors remain unaddressed.

There remains a need therefore for a novel analyte- responsive polymer that (1) responds to the presence of an analyte by altering its ability to propagate light and/or acoustic energy, (2) exhibits a specificity for a particular analyte, and (3) retains the ability to form a composite with the surface of a sensor such that the changes in the propagational properties are detected by the sensor.

The subject matter of the invention includes as one aspect a method relying on the acoustic and/or light exiergy propagational properties of an analyte-responsive polymer for detecting an analyte. The method comprises:

(a) contacting either an acoustic or optical detection system with an analyte to which an analyte-responsive polymer contained in the detection system is responsive;
(b) in the case of an optical embodiment of the invention, interfacing the system of step a with a means to detect changes in the light propagation of the analyte-responsive polymer;
(c) measuring propagational changes in the properties of the detection system; and
(d) correlating the propagational changes measured in step c with the presence, concentration, or rate of production of the analyte.

Another aspect of the invention is a composition of matter comprising a transducing polymer having from 1-20 crosslinks per polymer molecule and selected from the group consisting of amphOteric co- or terpolymers of pH between 5.0 to 8.0 of acrylic acid, alkyl methacrylate, and N, N-dimethyl-aminoethyl methacrylate, the amphOteric co- or terpolymer immobilized on a surface. A further aspect of the invention is the method of making the composition claimed.

Detailed Description of the Invention

Applicants have invented a method of detecting specific analytes in solution via a mechanism that makes use of the inherent propagational properties of crosslinked polymer films.

The precise nature of the changes in the propagational properties is not fully understood, but it is tentatively attributed to changes in the visco- elasticity, rigidity and acoustic and light wave propagational abilities of the analyte-responsive polymer film. Although changes in mass may occur, it is understood that the invention does not measure a wei xtt gain. Rather, the invention hinges on the resulting polymer-analyte complex having propagational properties different from that of the original transducing polymer.

Substitute Sheet

These propagational changes are detected by sensors with which the polymer-analyte complex is associated. The changes may be quantitatively correlated to the amount of analyte present. The inventive subject matter also includes unique compositions of crosslinked polymers immobilized on the surface of a sensor. The polymers may or may not contain receptors specific to particular analytes, such as ions, ligands, antibodies, etc. In the specific case of detection of hydrogen ions, impedance analyses of certain insoluble films of a crosslinked analyte-responsive polymer indicate changes in the polymer film properties when the pH of the medium is changed. When such a polymer is immobilized on the surface of a quartz microbalance, the resonant frequency of the polymer/quartz composite resonator decreases significantly in the isoelectric region. The resonant equivalent resistance and bandwidth increases upon approaching the pi, but decreases at the pi to values near those observed for the ionic forms. At pH extremes where the polymer is either fully protonated (cationic) or fully deprotonated (anionic), the polymer is swollen due to electrostatic repulsion between sites with similar charges. Conversely, at the pi the polymer is less swollen due to electrostatic crosslinking in the films. The changes in swelling and the attendant changes in viscoelastic properties and acoustic and light wave propagation produce unexpectedly large frequency changes. The invention describes two approaches to detecting changes that occur in a thin polymer film due to the influence of an analyte, in which the polymer film is coated on either a piezoelectric acoustic wave device or on a device capable of measuring changes in refractive indices of the polymer film. The principle of transduction for the two approaches is similar in that both rely on analyte-induced changes in the propagation of waves in the polymer film. This can be illustrated with a thickness shear mode acoustic wave device, for example an AT-cut quartz crystal coated with the polymer film, the combination thereof herein referred to as the composite resonator. The acoustic wave propagates back- and-forth through the thickness of the quartz crystal, across the polymer film-quartz interface and through the polymer f:.lm. This results in the establishment of a standing wave in the composite resonator with a frequency corresponding to the resonant frequency. The resonant frequency is determined by the thickness of the composite resonator and the propagation velocity of the acoustic wave. The energy attenuation of the acoustic wave is affected by viscous losses encountered by the acoustic wave during propagation. Since the thickness, propagation velocity, and energy attenuation in the quartz crystal are not affected by the analyte, the composite resonator is sensitive only to analyte-induced changes in the properties occurring in the polymer film. The resonant frequency is therefore affected by analyte- induced propagational changes and the modulus of the polymer film, and the energy attenuation by viscous losses in the polymer film. That is, the energy attenuation depends upon the amount of energy "reflected" back into the composite resonator during oscillation. This is customarily defined as the quality factor Q, which is the ratio of energy stored to energy lost during oscillation of the resonator. Increasing energy attenuation results in lower values of Q, higher values of the equivalent resistance R and increases in the bandwidth of the electrical conductance.

The optical sensing approach is similar in that light waves propagate back-and-forth across the polymer film interfaced with a support and into the polymer. In essence, a standing wave is established in the composite optic sensor. The standing wave may be a single frequency, or more commonly, a range of frequencies. The amount of light reflected back into the detector depends upon the refractive index of the polymer film. Since the refractive index of the support is independent of the analyte, changes in the refractive index of the polymer film due to the influence of analyte are responsible for changes in the amount of light reflected back and ultimately detected by the optic sensor. The resulting changes in the light intensity are therefore analogous to the changes in the energy attenuation in the acoustical wave device. The two devices are related conceptually in that both are sensitive to changes in the transmission of wave energy resulting from the influence of the analyte on the polymer film properties. In addition, it is feasible that specific wavelengths of light can be absorbed by the polymer film, and the degree of absorption can be influenced by the analyte interaction with the polymer. This can result in changes in the intensities of different wavelengths, therefore changing the weighted average frequency of light exiting the sensor. This is analogous to the changes in the resonant frequency detected in the acoustic wave device. Both devices therefore are similar conceptually when described according to principles of wave propagation.

In support of the disclosure of this invention, the following terms are intended to convey the following meanings.

"Analyte" ("A") means any substance capable of interaction or reaction with an analyte-responsive polymer, including analyte-responsive polymers containing Biological Receptors. The Analyte may be formed or consumed by chemical or biological processes, including the growth and metabolic processes of cells, microorganisms or their subcellular components. The Analyte may be free-floating in a media, attached to a cell surface, contained within a cell, or produced on the surface of the analyte-responsive polymer as a result of enzyme catalysis. The Analyte alternatively may be the product of catalytic reactions, enzyme immunoassays, or DNA probe assays. The Analyte may be an acid, buffering agent, salt, enzyme, protein, carbohydrate, lipid or similar biological product or a substance that is diminished or increased in concentration by biological activity.

"Biological Receptors" are substances that form specific binding pairs with the Analyte. These may variously be chelating agents, antibodies, lectins, tissue receptors, cellular adhesion factors, ligand binding proteins and similar analyte receptor reagent substances. "Transducing Polymer" means any polymer that, as the result of interaction with an Analyte, exhibits a change in the properties governing its behavior in the propagation of acoustic or light energy. The transducing polymer may or may not be amphoteric. "Analyte-Responsive Polymer" ("ARP") means a crosslinked amphoteric polymer that 1) is capable of selective interaction or reaction with an Analyte, and 2) exhibits a change in the properties governing its behavior in the propagation of acoustic or light energy as the result of interaction with an Analyte. The ARP may contain a Biological Receptor capable of forming a specific binding pair with an Analyte.

The polymer is chosen or designed to react with a analyte through a variety of reactions including ion pairing, complexation reactions, redox reactions, or covalent coupling. This adaptability enables the invention to apply equally well to large or small molecular weight analytes.

"Amphoteric Polymer" means an analyte-responsive polymer (either natural or synthetic) which contains both acidic and basic groups. Amino acids and proteins are amphoteric since they contain both acid (-COOH) and basic (-NR_2) groups.

"Polymer-Analyte Complex" ("C") means a complex formed upon the reaction or interaction of an Analyte with an Analyte-Responsive Polymer or Amphoteric Polymer. For convenience, "complex" as used herein refers to the substance resulting from the reaction of a analyte-responsive polymer with an analyte regardless of the the specific mechanism involved.

"Propagational Changes" means changes in properties of the analyte-responsive polymer as a result of the polymer's complexation or reaction with an Analyte which may involve, as compared to that of the polymer alone, 1) changes in its ability to propagate either acoustic or light energy or 2) changes in the rigidity, elasticity and/or the viscosity of the polymer-analyte complex, or 3) any alteration in the phase composition of the polymer. The propagational changes in the polymer-analyte complex may be heterogeneously distributed in micro-domains on or within the attached polymer or may be homogeneously distributed on or in the polymer matrix.

"Sensor" ("S") refers to a device to detect propagational changes in the polymer coating material. Piezoelectric oscillator, quartz crystal microbalance and QCM are names for devices that use piezoelectric principles as the basis for detecting such changes. In addition, shear horizontal acoustic plate mode devices (SHAPM) are alternative devices for use in detecting these changes. A waveguide optical biosensor (WB) can be used to detect changes in light propagation properties of the complex. The transducing polymer is applicable to both planar and fiber (cylindrical) wave guide formats including fiber optics, interferometers, refractometers, Mach-Zender and optical grading devices.

"QCM" refers to a bulk acoustic wave device operating in the shear mode typically comprising either AT or BT cut quartz where the quartz is sandwiched between two excitation electrodes.

"Impedance Analysis" refers to any analysis technique that measures the current across the surface of a quartz crystal at a constant voltage over a specified range of frequencies. "Composite Sensor" refers to the unit of an analyte-responsive polymer film bonded to the surface of a sensor.

"Detection System" refers to the unit of the analyte-responsive polymer, sensor, and test medium. "Organism" is meant to include any organism which, as a result of its metabolism, makes a product unique to that organism that can be detected or identified by the method of this invention. The organisms for which this invention will be most useful, however, are micro- organisms normally grown in aqueous cultures, such as bacteria, fungi, and tissue cells.

"Growth Regulators" are substances that stimulate or retard the growth of the organism.

"Nutrients" are substances metabolized by the organism and necessary for its growth.

The term "XAMA-7®" will refer to any pure chemical composition of pentaerythritol-tris-(B- aziridinyl)propionate. XAMA-7® is a registered trademark of the Virginia Chemical Co.

Substitute Sheet

The crosslinked analyte-responsive polymer (ARP) contains analyte-sensitive moieties and is attached to the surface of a sensor forming a Composite Sensor. A sample of test medium suspected of containing an Analyte is brought into contact with the sensor and any Analyte present interacts or reacts with the ARP to form a polymer-analyte complex. A propagational change is detected in the complex as compared to that of the uncomplexed ARP. The change is manifested as a change in resonance frequency of a piezoelectric oscillator or as a change in light propagation properties by a waveguide optical sensor or other optical detection device.

The invention is particularly useful for the monitoring of biological systems to detect determine changes in the concentrations of a variety of substances including cellular metabolites, the products of enzymatic reactions, pharmaceutical compositions, industrial chemicals, or any product of a biological system that requires continuous real-time (kinetic) measurement. This invention is also capable of detecting any organism which, as a result of its metabolism, makes or uses a product characteristic of that organism that can be detected or identified by the method of this invention. The invention can be used to detect organisms normally grown in aqueous cultures, such as bacteria, fungi, and tissue cells. [v] However, it is not necessary that these organisms remain intact. The invention is intended to operate as well with disrupted or solubilized components of the organism. A more specific illustration of the invention is the use of a polymer designed to react with analytes produced as a result of pH changes.

An example of such an analyte is an H^+ group. When metabolic organic acids or carbon dioxide produced by organisms acidify the growth medium, protons (analytes) released as the metabolic products react with proton receptor groups on the analyte-responsive polymer, thereby altering the acoustic propagation properties of the polymer. The changes in propagation occur as the pH of the medium approaches the isoelectric point of the complex. This propagational change in the ARP effects a change in the resonant frequency of the piezoelectric oscillator that can be read electronically to determine the metabolic rate and cell growth rate of a culture. Alternatively, reaction of the protons with the polymer can be detected by a waveguide optical sensor or other optical detection device as a change in light propagation properties. The selectivity of the polymer/analyte interaction also can be controlled by the design of the polymer. For example, antibodies, polynucleic acids, receptors, chelating agents, cellular adhesion factors, and ligand binding molecules can be linked to the polymer. By varying the composition of these receptor sites on the polymer, the invention can be made highly selective for a specific analyte or broadly responsive to a number of analytes.

A more specific illustration demonstrates the analyte selectivity of the invention. A good example is an antigen which complexes with an ARP constructed to contain the complementary antibody as a Biological Receptor. Changes in the amount of crosslinking of the resulting antigen/antibody-polymer complex induce a propagational change in the ARP and produce a change in the resonant frequency of the piezoelectric biosensor. As in other examples, monitoring of the resonant frequency of the piezoelectric oscillator provides a specific measurement of the amount of metabolite or the rate of metabolite production in the medium.

Measurements may also be made by means of a light propagational or optical sensor.

Piezoelectric Oscillator

The ability of piezoelectric transducers to detect small changes in mass, viscosity and density at their surfaces while immersed in liquids is well known in the art (Ward *et al.* Science (1990) 249:1000-1007 and Frye *et al.*, Appl. Spectroscop. Rev. (1991) 26:73). In addition, shear surface acoustic, flexure and shear-horizontal acoustic plate mode devices can be appropriately modified for use as sensors (Lu and Czanderna, Eds., Elsevier, New York, (1984) pp. 351-388 and Guilbault et al., CRC Crit. Rev. Anal. Chem. (1988) 19:1, and Wohltjen et al., ACS Symp. Ser. (1989) 403:157). Most applicable to the present invention is the shear mode AT-cut quartz resonator, commonly referred to as the quartz microbalance (QCM), which comprises an AT-cut quartz crystal sandwiched between two metal excitation electrodes which generate a standing shear wave across the thickness of the quartz crystal. The shear wave experiences an antinode at the surface of the quartz crystal and propagates into the film on the surface of the crystal; the thickness of the film and the nature of the shear wave propagation in the film determine the frequency response. The frequency response of the QCM generally is interpreted in terms of a mass increase on the resonator surface, which causes a corresponding decrease in the resonant frequency according to the Sauerbrey relationship (Sauerbrey, Phys. (1959) 155:206) (eq. 1), where Δf is the measured frequency shift of the initial (resonant) frequency (f) of the quartz crystal, Ãm is the mass change, A is the piezoelectrically active area defined by the two excitation electrodes, Pq is the density of quartz (2.648 g/cm^{-3}) and Uq is the shear modulus (2. 947 × 10^{11} dynes cm^{-2} for AT-cut quartz).

$$2f_0^2 \Delta m \text{ (eq. 1) } \Delta f = A\ Pq\ Uq$$

Surface acoustic wave (SAW) and shear horizontal acoustic plate mode (SH-APM) devices represent an alternative piezoelectric transduction techniques applicable to this invention. These devices comprise interdigitated microelectrode arrays on the surface of a piezoelectric quartz substrate. They exhibit frequency changes that can be correlated with mass changes or stiffness coefficient at their surface arising from changes in the velocity of a transverse surface wave. These devices have also been employed as viscosity sensors.

Optical Sensors The use of optical biosensors to detect the presence of various analytes is common and many examples may be found in the art. (Place et al., Optical- Electronic Immunosensor: "A Review of Optical Immunoassay at Continuous Surfaces", Biosensors, 1, 321- 353). Suitable types of optical sensor devices include planar waveguides (Burgess, Proc. SPIE-Int. Soc. Opt. Eng., 1368 (Chem., Biochem., Environ. Fiber Sens. 2), 224-9 (1991)), optical fibers (Bluestein et al., "Fiber Optic Evanescent Wave Sensors for Medical Diagnostics", TIBTÎCH, £, 161168 (1990)), metalized prisms (Kooyman et al., "Surface Plasmon Resonance Immunosensors", Analytical Chem. Acta., 212., 35-45 (1988)) and diffraction gratings. The detection surfaces of optical sensors can be planar or cylindrical (fibers). Generally, optical biosensors respond to small changes in the index of refraction at the surface of a waveguide. This change in the index of refraction results from the selective binding of an analyte to an immobilized analyte-receptor on the surface of the optical sensor. Light traveling in the waveguide induces an evanescent wave in the test media above the waveguide. The interaction of this evanescent wave with analyte/receptor complexes alters either the intensity of the evanescent wave or the coherency of light propagated in the waveguide. Evanescent wave intensity can be altered by changes in fluorescence, adsorption, or light scattering properties of the resulting analyte/receptor complexes. In this way the formation of an analyte/receptor complex can alter the light intensity. Alternatively, the phase or coherency of light traveling through single mode waveguides can be altered by the analyte/receptor interaction and the changes in phase or coherency resulting from the analyte/receptor interaction measured by interferometric devices (e.g., Mach-Zehnder).

Polymers used in preparation of an analyte- responsive polymer matrix are polyampholytes containing both acid and base functionalities and having a defined isoelectric point (pi). Polyampholytes of both synthetic and biological origins can be used and can be comprised of both synthetic (e.g., acrylic, etc.) or biological (e.g., aminoacids, etc.) monomers. The polyampholytes have a molecular weight generally in the range of 500 to 500,000 and more preferably, in the range of 1000 to 100,000. The analyte specific polymer is formed by crosslinking the polyampholyes, forming an analyte responsive polymer matrix having a crosslinked density of 1 to 20 cross-linkers per polymer molecule and having a thickness ranging from monomolecular to several microns, and more preferably from 0.05 to 5 microns.

Specific polymers used as analyte-responsive polymers for the examples of a sensing device using acoustic or light energy are co- or terpolymers of acrylic acid (AA), alkyl methacrylate (RMA), and N, N-dimethylaminoethyl methacrylate (DMAEMA). Structurally, the polymers may be linear or may contain pendent or crosslinking chains. They are prepared using a two-step process outlined in U.S. Patent 4,749,762 herein incorporated by reference.

The first step produces a prepolymer from methyl acrylate (MA), RMA and DMAEMA. The second step is the controlled selective hydrolysis of methyl acrylate segments to form a product with pendant acid and base groups. Figure 8.1 illustrates the two-step polyampholyte preparation process. Other synthetic methods, including Group Transfer Polymerization (GTP), may also be incorporated in preparation of amphoteric polymers. A two-step emulsion process is preferred over direct solution polymerization for the following reasons. Michael addition of the amine to acrylic acid monomer is more easily prevented. Emulsion polymerization of prepolymer is significantly faster and more easily controlled than solution polymerization. In addition, the composition distribution of monomer units in the polymer is far more easily regulated by controlled feed methods. The polymers are generally soluble in water at all pH's other than their isoelectric points (pi). The isoelectric point is determined by the ratio of acid to base groups and thus can be varied by synthesizing polymers with appropriate ratios. The preferred ratio for monomers used in the instant invention is such that the polymer's pi is in the physiological region (pH -6.2-8.0). This normally

corresponds to a ratio of acid to base of 2.0 to 0.8 for acrylic acid and DMAEMA. (The pi is necessarily dependent on the pKa values for the component groups.) Responses can also be affected by the nature and size of the neutral, non-ionic segment, which can influence the pK of the component acid and base moieties. The solubility characteristics of the polymers are strongly influenced by their ion content. Polymers having significant neutral hydrocarbon segments are less water- soluble at their isoelectric point than polymers with few or no neutral segments. Applicants have prepared a series of polymers containing different alkyl methacrylates and having variable segment fractions, all with pi's in the physiological range. Several other polymers having different solubility and associative characteristics are also predicted to be suitable.

The sensitivity of the polymer to small changes in pH is largely dependent on the narrowness of its composition distribution. This is represented in Fig. 8.2. A narrow composition distribution is generated by controlling the ratio of reacting monomer in the reaction medium. This ratio is not that found in the polymer but is determined by the reactivity ratios of the constituent monomers. The ratio can be maintained by using either a balanced feed or a starved feed reaction process. The balanced feed process described in the U.S. Patents 4,735,887 and 4,749,762, incorporated herein by reference, requires careful reaction control, but leads to rapid formation of high molecular weight product. The starved feed process is preferable when rapid production of high molecular weight product is unnecessary. The starved feed process involves the addition of feed monomer at a rate much lower than its bulk reaction rate in neat media. The reaction becomes essentially a living free radical process, occurring only when monomer encounters an emulsion particle containing a living radical. Enough "balance monomer" is added to saturate the aqueous phase (determined as the point where the solution starts to develop translucence), before starting addition of initiator. A slight excess of MA should also be maintained to give the correct product composition. This is easily accomplished because of the favorable relationship between total inherent reaction rate and reactivity ratios for the acrylate-methacrylate system. Many combinations of monomer are capable of yielding polymers having pi's in the physiological pH range. Amphoteric polymers can be prepared from various combinations of the following sets of monomers set out below:

(A) *Acidic Monomers*: Molecular or ionic substance that can yield a hydrogen ion to form a new substance. Examples are acrylic acid, methacrylic acid, and monomers containing phosphoric acid and sulfinic acid groups.

(B) *Basic Monomers*: Molecular or ionic substance that can combine with a hydrogen ion to form a new compound. Examples are DMAEMA, diethylaninoethyl methacrylate, t-butylamino- ethyl methacrylate, morpholinoethyl methacrylate, piperidinoethyl methacrylate.

(C) *Neutral Monomers*: Molecular or ionic substance that is neither acidic or basic. Examples are alkyl methacrylates (methyl MA, ethyl MA, butyl MA), hydroxyethyl methacrylate (HEMA), hydroxypropyl methacrylate, vinyl pyrrolidone, vinyl acetate (vinyl alcohol on hydrolysis), acrylamides, vinyl ethers, styrene. (Reaction of co-monomers having vastly differing reactivity ratios requires very careful reaction control and therefore is not preferred.) In addition to the examples set out above and in Table A, any aqueous soluble amphoteric polymer with a pi in the physiological range may be useful as a pH sensitive analyte-responsive polymer. Specific examples include:

1. Polymers generated by the reaction of dimethylamino ethanol and similar compounds with methylvinylether/maleic anhydride co-polymers.
2. Hydrolyzed co-polymers of vinyl pyridine and methyl acrylate.

Optionally, the analyte responsive polymer may contain an analyte receptor reagent capable of selective analyte binding. The receptor reagents can be chemically linked to pendent functional groups (such as hydroxyl, carboxyl, amino, thiol, aldehyde, anhydride, imide and epoxy) on the polymer or immobolized within the analyte responsive layer by means of entrapment within the polymer matrix. Alternatively, the analyte receptor reagents could be immobolized by adsorptive interactions with the polymer or a polymer matrix component. This forms an intermediate group that is reactive towards the analyte. In the assay, the analyte attaches to the activated polymer by reaction with the intermediate coupling group.

A wide variety of analyte receptor reagents are contemplated. These include one member of an analyte specific binding pair. Members of specific binding pairs may be of the immune or nonimmune types. The immune types include antibodies, whether polyclonal, monoclonal or an immunoreactive fragment such as Fab- type, which are defined as fragments devoid of the Fc portion of the antibody (e.g. Fab, Fab' and $f(ab')_2$ fragments or so-called "half-molecule" fragments obtained by reductive cleavage of the disulfied bonds connecting the heavy chains). If the antigen member of the specific binding pair is not immunogenic (e.g., a hapten) it can be covalently linked to a carrier protein to render it immunogenic. Polynucleic acids and receptors are also contemplated as analyte receptor reagents.

Non-immune binding pair members include systems wherein the two components share a natural affinity for each other but are not antibodies. Exemplary non-immune analyte receptor reagents include avidin, streptavidin, complementary probe nucleic acids, binding proteins, chelation reagents, cellular adhesion factors and ligand binding proteins.

Linking chemistry for the attachment of analyte receptor reagents involving gluteraldehyde, cyanogen bromide, hydrazine, bisepoxirane, divinylsulfone, epichlorohydrin, periodate, trichlorotriazine, diazonium salts, carbonyldiimidazole, carbodimides, N-hydroxy succinimide, and tosylates has been described extensively in the art. A review of these chemistries and procedures is given in "Practical Guide for Use in Affinity Chromatograph and Related Techniques", Reactifs IBF—Societe Chimique Pointet-Girard, Villeneuve-La- Garenne—France. Specific examples of the activation of pendant hydroxy groups by carbonyldiimidazol are reported in J. Biol Chem (1979) 254:2572 and J. Chromatogr, (1981) 219:353-361. The activation of pendent carboxylic acid groups by water soluble carbodiimide is described in Biochem J. (1981) 199:297-419. Activation of carboxylic acids by N-hydroxysuccinimide is described in Biochemistry (1972) 111:2291 and Biophys. Acta (1981) 670:163. Typical water-soluble polymers that may be activated by these procedures may include polyhydroxy- ethyl methacrylate and acrylate, methylvinylether copolymers, polyvinyl alcohols and copolymers, and the polyampholytes described above. Other water-soluble polymers may also be used.

Polymer Crosslinking The analyte-responsive polymers of the present invention may be crosslinked by several mechanisms where the carboxylic acid moiety is the reactive site. Crosslinking may occur by reaction of pendant carboxylic acid groups with multifunctional aziridines or they may be crosslinked through pendant carboxylic acid groups by reaction with multifunctional epoxides. This reaction is typified by the crosslinking of 1,4-butanediol diglycidyl ether where the ring opening mechanism is similar to that of the reaction of aziridines. Alternatively, pendant primary amine groups can be formed on the analyte-responsive polymers by reaction of a few carboxylic acid groups with propyleneimine or ethyleneimine, which will allow the polymers to be crosslinked by reaction with a number of crosslinkers normally used in photographic systems for crosslinking gelatin such as aldehydes, carbodiimides, and others. A number of crosslinkers and crosslinking reactions used to crosslink gelatin, also capable of crosslinking the analyte-responsive polymers of the present invention through either acid or primary amine functionality are described in Pouradier and Burness (In the Theory of The Photographic Process, 3rd ed., C.E.K. Mees, Ed., p. 54-60), herein incorporated by reference. Less commonly used but equally applicable are methods involving ionic crosslinking through metal ion coordination and reaction of ionic clusters, as well as various types of covalent bonding involving reaction of aziridines and Michael addition of olefins. For the analyte-responsive polymers of the present invention crosslinking or the pendant carboxylic acid groups with multifunctional aziridines is most preferred.

Growth Media The basic growth medium for the invention contains nutrient materials selected for fermentation by the specific type cells to be analyzed. A specific analyte- responsive polymer (ARP) of the type described above is non-reactive with the medium alone. The medium can be optionally supplemented with growth regulators such as antibiotics, amino acids, vitamins, salts, or lipids. The growth regulators, such as hormones, used separately or in combination, may be used to make the medium either highly selective for a specific cell type or less selective for a broad spectrum.cell response. The composition of the nutrient medium enables the invention to respond flexibly to a variety of analytical needs. Components for growth media are commercially available from Difco. (Detroit, Michigan) and BBL (Coclceysville, Maryland) among other sources. A detailed description of the growth media needed to practice the instant invention is reviewed by Ebersole *et al.* (Int'l. Pub. No. WO 91/01381) herein incorporated by reference.

Before testing or following sterilization, the pH of the nutrient medium was adjusted as needed to a value compatible with the physiological requirements of the cell culture and the propagational stability of the polymer.

Cell Concentration Determination

Cell concentration in a given experiment was estimated microscopically in a Petroff-Hauser counting chamber. Growth of the cell culture was first arrested by the addition of 0.1 per cent sodium azide solution. In some cases, the cell densities in the Petroff-Hauser counting chamber were counted automatic—ly using an Olympus Q2 image analyzer equipped with a, .ampaq 386/25 computer.

Piezoelectric Biosensor Dev ce and System Cell detection and identification, antibiotic response studies, and growth rate determinations can be made either with a single piezoelectric oscillator device under the conditions outlined above or with a piezoelectric oscillator device referenced against a second quartz crystal that compensates for instabilities associated with temperature. Such a device is described by Ebersole et al. (Int'l. Pub. No. WO 91/01381) and is herein incorporated by reference.

Use of Impedance Analysis

The interaction of the polymer with the surface of the QCM is mechanical in nature and may be analyzed by impedance analysis. Impedance analysis measures the current across the quartz crystal at a constant voltage over a specified range of frequencies. Impedance analysis was performed with a Hewlett-Packard 4194A Impedance/Gain-Phase Analyzer (HPIB) capable of performing measurements over a frequency range of 100 Hz-40 MHz in the impedance mode. Data collection was accomplished via an HPIB interface with a Macintosh personal computer. Impedance analysis is a technique well known to those skilled in the art and is outlined in Mura atsu et al., Anal. Chem., 60:2142 (1988) herein incorporated by reference.

Measurements of Refractive Index

An alternative detection system to acoustic sensors are optical devices capable of detecting changes in the refractive indices of polymer films. In the present invention analyte-receptor binding will produce small changes in effective index of refraction (n_{eff}) of the analyte- responsive polymer immobilized on a surface. The most preferred method of refractive indices measurement makes use of optical fibers and a single mode Mach-Zehnder interferometer in which the guided light is split into two parallel arms. One arm of the interferometer is coated with an analyte receptor while the other arm is protected to provide a reference path. As a result, light conducted into the waveguide is split into two beams. When analyte binding occurs, the refractive index at the surface of the receptor coated arm is altered whereas the effective index of the second beam (reference arm) does not change. When light in the interferometer arms are recombined, constructive or destructive interference can occur. If the arms of the interferometer are the same length (LI = L2 = L), the difference in light propagation (ΔΦ) resulting from analyte binding can be mathematically described by equation 2, where λ is the wavelength of input light, L is the path length, and nl and n2 are the index of- the reference and test- arms respectively.

$$2\ \Delta\Phi = —\ 2\Pi—L\ (n2–nl)\lambda$$

The phase difference is directly proportional to the difference in effective index difference (n2-nl) of the waveguide legs. Since the extent of analyte binding affects the $n_e f_f$ = (n2-nl), the output intensity of the interferometer can be related to index changes resulting from the refraction induced by analyte binding. Devices that measure the changes in refractive indices of an analyte-responsive polymer film provide versatile transduction methods capable of detecting both direct analyte binding events as well as enzyme amplified analyte binding.

Examples

The following non-limiting examples illustrate the basic principles and unique advantages of the present invention.

Example 1

Procedure for Crosslinking Polyampholytes with Multi-functional Aziridines and Pemonstration of pH-cc-ntrolle, pys Binding Test solutions of polya pholyte and multi¬ functional aziridine crosslinker (ama-7®) having ratios of 1:1, 5:1, 10:1, 50:1 and 100:1 polymer acid to aziridine were prepared by mixing stock solutions of 0.154 Normal (1-1-1) AA-MMA-DMAEMA polyampholyte (pi-7.0 in 20% methanol/water acidified to pH = 6.0 with HC1) and 1.405 Normal Xama-7®. Before mixing the stock solutions, sufficient ammonium hydroxide was added to the polymer solution to shift the pH above 11.0 to prevent premature reaction of the aziridine with the polymer acid. The mixtures were then spin coated onto glass plates and piezoelectric crystals at rotor speeds of 1000, 2000 and 4000 rpm. The coated materials were then placed into a circulating air oven for 5 min at 100°C to drive off the ammonia and cure the composition. The plates were washed in water and tested for physical integrity. Coatings formed from the 1:1 and 5:1 mixtures were hard and essentially unswollen. The 10:1 and 50:1 compositions gave insoluble but swollen gel coatings with good adhesion to support, physical integrity and robustness. The 100:1 gel showed poor integrity and was easily abraded from the support. Therefore, the 10:1 and 50:1 compositions were chosen for further study.

The thickness of 50:1 coatings, as measured by elipsometry, were 0.2 ìm,

0.4 ìm and 0.8 ìm for spin castings made at 4000, 2000, and 1000 rpm respectively. The charge on the surface of each film was demonstrated by placing a series of buffers, each containing an identical anionic blue dye, on each coated composition. After allowing the compositions to sit for 30 min, the compositions were washed with neutral water to remove unattached or unabsorbed dye. All dye spots corresponding to buffers with pK <7.0 remained easily detectable while all spots corresponding to buffers having pK >7.0 washed immediately from the coating. This was in spite of the fact that the basic buffers swelled the compositions as well as the acidic buffers. This demonstrates the pH-controlled dye binding characteristic of the polyampholyte coating. The coated piezoelectric crystals prepared as described above were tested for microbial and enzymatic response and used in the following Examples concerning detection by piezoelectric means.

Example 2: Polymer Preparation and Immobilization

The polymer (1) of Fig. 8.3 was synthesized according to the procedure of outline by Foss (U.S. Patent 4,749,762) and by the following procedure. Under a nitrogen atmosphere, emulsifier solution (1000 mL distilled water, 10 g Triton QS-30 surfactant (Rohm & Haas, Philadelphia, PA) and 10 g N, N-dimethylamino- ethanol) was heated to 60°C. Then a mixture containing 53 mL of methyl acrylate, 53 mL of methyl methacrylate, and 98 mL of N, N-dimethylaminoethyl methacrylate was added at 4 mL/min. When the solution became saturated (saturation is evident when the solution begins to exhibit translucency), an initiator solution (5 g of ammonium persulfate in 250 mL distilled water) was added at 0.75 mL/min while simultaneously continuing the momomer addition. Addition of the initiator solution resulted in immediate polymerization as evidenced by a temperature increas. After all the monomer was added, the mixture was stirred for an additional 15 min, and was then poured into a 2 L polyethylene beaker. Acetone was added until the product coagulated. The product was then collected by filtration and washed with water to remove remaining emulsifier and other impurities. The polymer was transferred to a flask equipped with a high shear blade stirrer, 800 mL ethanol was added and the µm mixture heated to 80°C. After the polymer had dissolved, a solution of 32.65 g KOH in 100 mL distilled water was added via an addition funnel in order to selectively hydrolyze the methyl methacrylate to the acrylic acid salt. The addition rate was controlled so that the polymer did not precipitate during this step.

After the addition was complete, the mixture was stirred for an additional 30 min at 80°C. The product was isolated and purified by isoelectric precipitation. This was done by transferring the ethanolic polymer solution into a large excess of distilled water and then adding an equivalent amount of hydrochloric acid to shift the solution pH to the polymer's isoelectric point (pi). The polymer was then isolated by centrifugation, washed with water buffered at the pi and redissolved in water containing a small amount of ethanol at a pH either above or below the pi. The polymer was stored in this slightly acidic or basic solution because if isolated and dried at its pi, redissolving was slow and difficult. Each side of an AT-cut quartz crystal was coated with 2000 A thick gold electrodes. Underlayers of 500 A thick titanium were used for adhesion in the center of the quartz crystal. One side of the crystal was spin-coated with the crosslinking polymer solution at 1000 rpm for 40 sec. The composite resonator thus formed was allowed to dry in air and was then heated at 100°C for 10 min. in a circulating air oven.

Example 3: Polvamoholvte Crosslinking

Films of polyampholyte (1) (Fig. 8.3) were crosslinked with a multifunctional aziridine according to the procedure depicted in Fig. 8.3. The crosslinker, pentaerythritol-tris-(B-aziridinyl)propionate (3) (Xama- 7®), is prepared by the Michael addition of ethylene imine to pentaerythritol triacrylate (2). Crosslinked films of polyampholyte (1) were then prepared by spin- coating onto a support a solution containing 5. 4^g of polyampholyte (1) and a required amount of Composition (3), made basic with ammonium hydroxide to block premature crosslinking. The films were then crosslinked in place by mild heating for 5 min. at 100°C. It was found that generally a 50:1 equivalent ratio of 1:3 gav best results. The heating step eliminates NH_3, thereb. allowing the coating to become mildly acidic thus enabling crosslinking of the polymer via a proton- assisted ring-opening reaction between the aziridine groups of the crosslinker and a small number of -CO_2H groups of polyampholyte (1). The film thickness was controlled by spin-coating rotational speeds, which ranged from 1000-4000 rpm. Film thicknesses were determined independently with a Sloan Dektat IIA stylus profilometer.

Example 4: Piezoelectric ARP-Polv er Response to PH Change

This example illustrates that changes in pH of the test medium alter the properties of the crosslinked ARP. The resulting changes can be detected as alterations of ARP thickness, contact angle, and viscoelastic properties as reflected by alterations in piezoelectric oscillator response. Measurements of pH were made simultaneously by frequency measurements or network analysis and a Beckman Model 032 pH meter. The analog output of the pH meter was connected to an IO/Tech analog-digital, converter interfaced to the HPIB, enabling automatic measurement of pH. Sessile contact angle measurements were performed with a Rame-Hart contact angle goniometer using a 10 ìL water droplet.

Film thicknesses in aqueous solutions were measured directly with a phase measurement interferometric microscopy (PMIM) (Zygo, Inc.). Crosslinked films were coated on evaporated gold films on quartz substrates according to the procedure described in Example 3 and were immersed under a thin film (ca. 1 mm) of water that was contained with a glass cover slip. The pH of the solution was changed by replacing the water between the sample and the cover slip- with water adjusted to-the desirable pH value. A reference height difference for calibration was provided by the gold films, whose thickness was established from the frequency shift of the quartz crystal microbalance during electron beam vaporation. In this example, the pH of the test mileu covering QCMs on which a 1:1:1 AA-MMA-DMAEMA crosslinked ARP immobilized as described above, was acidified with 0.001 N KC1 over the pH range of 9.0 to 3.0. Films of thicknesses of 0.8 ìm and 0.4 ìm respectively of were analyzed. The series resonant frequency (fs) measured with the composite resonator in the feedback loop of a broad-band amplifier, and the frequency of maximum conductance ($f_G$max) measured by impedance analysis, were found to change significantly when the pH of the solution exposed to the resonator was altered. (It is assumed throughout that the difference between fs and f<$_{3m}$ax is negligible). When the pH of the medium was gradually increased from pH = 3.0, an abrupt frequency decrease was observed at pH = 4.8 (Figure 8.4a and 8.4b). The magnitude of the frequency shift increased with polymer film thickness, giving exceptionally large shifts approaching -6000 Hz fc 0.8 ì thick polymer film. The resonant frequency exhibited a slight increase at pH = 5.5, followed by an abrupt increase at pH >6.1. The increase in the center of the isoelectric region pH = 5.5 was more apparent for the thinner films (0.4 ìm), which also exhibited a broader isoelectric region based on the frequency changes. These data strongly suggest that the observed frequency changes were related to the changes accompanying the transitions between the ionic and isoelectric forms of the polymer. In addition, the frequency changes for the 0.4 and 0.8 ìm thick films correspond to mass changes of approximately 20 ìg cm"^{2} and 90 ìg cm"^{2} which is substantially larger than the total areal mass (Äm/A) of these films after spin costing.

Example 5

Urease Activity Piezoelectric Measurement bv ARP-OCM The response to pH changes induced by urease- catalyzed hydrolysis of urea was measured in 2 mL of a buffer solution in which the resonator coated with crosslinked polymer 1 (Figure 8.3) was immersed. The buffer solution was prepared from 1.48 mL 1.0 mM NaOH, 8 mL 0.2 mM EDTA and 100 ð. of deionized water adjusted to a pH of 5.5 with phosphoric acid. Measurements of the pH changes dependent upon urease concentration were performed by adding known amounts of a urease solution (1 mg urease (Sigma, St. Louis, MO) in 100 mL of deionized water) to the buffer solution containing 0.25 M urea (Fisher Scientific Co., MO) while the resonator was immersed in the solution. Conversely, the response dependence upon urea was determined by adding known amounts of a urea solution to the buffer solution containing 0.1 ìg/mL urease.

Urease-catalyzed hydrolysis of urea results in the formation of NH_3 with a corresponding increase in the pH of the medium. Accordingly, when urea was added to a phosphate buffer solution (initial pH = 4.0) containing urease, a monotonic frequency decrease is observed after a short time followed by a monotonic increase until the original frequency was attained (Figure 8.5). The rates of frequency change in both branches were essentially identical. The time at which the frequency reached the minimum shifted to longer values with decreasing urease concentration. This data is consistent with pH-dependent frequency of the composite resonator: urease-catalyzed hydrolysis increases the pH of the medium resulting in conversion of the polymer/analyte complex to its isoelectric form where the frequency decreases. After the pH exceeded the pi the forequency increased once again.

9

Electronic and Photonic Molecular Materials and Devices

9.1 Electronic and Photonic Molecular Materials Group

Introduction

The Electronic and Photonics Molecular Materials group is based in the Department of Physics and Astronomy at the University of Sheffield.

We have a long-standing interest in many aspects of the physics and technology of organic semiconductors and organic optoelectronic devices. Current research programmes include study into organic light emitting diodes (LEDs), organic photovoltaic devices (PVs), organic field-effect transistors (FETs) and organic photonic devices.

Our work is conducted in a series of modern laboratories and clean rooms for device preparation and evaluation. Micro- and nano-scale devices and structures are also developed jointly with researchers in the III-V Central Facility at Sheffield. We have access to a range of spectroscopic techniques, including an ultra-fast laser facility and a range of scanning probe microscopes.

We have had a number of joint research projects, running with both UK and European Universities and also with UK and international companies commercializing organic electronics.

9.2 Organic Light Emitting Diodes

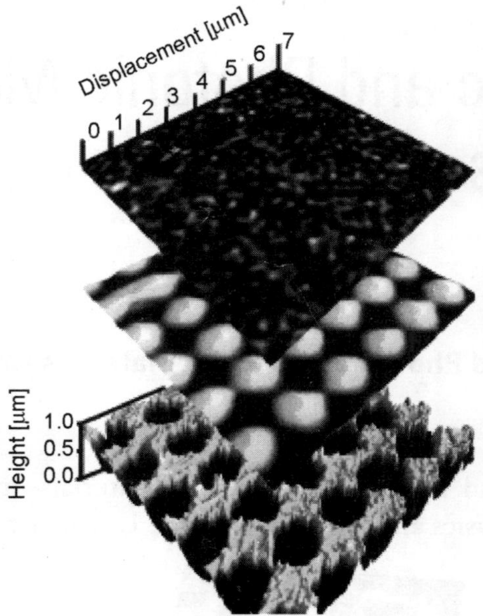

The group has more than a ten years' research experience in the development of polymer LEDs. Our work has focussed on developing optical methods to improve device external efficiency. This is important because in conventional polymer LED's only about 30% of the light ever escapes the device usefully, the remainder being trapped in 'waveguide modes'. We have developed a number of ways to defeat waveguide modes, including the development of new methods to micropattern polymer LED pixels using laser interference lithography. This process permits light trapped in the device to exit to the outside world.

Right: A sequence of images showing the preparation of a patterned anode. The pattern acts as an optical grating, and changes the 'in-plane wave-vector' so that totally internally reflected light can escape from the device.

We have also looked at using new high-reflectivity materials for the device cathode to reduce optical loss due to internal absorption of light and thereby increase external optical efficiency. Much of this work has been done in combination with CDT (the through the DTI-sponsored LEAP and MOET projects). An important aspect of our work in this field is the development of 'manufacturable' solutions in collaboration with existing industrial capabilities, allowing our industrial partners to improve the performance of their existing devices at low cost. It is hoped that the results of the MOET project will help to stimulate the use of polymer LEDs in lighting applications, where the power efficiency of LED devices offer important power savings, with manufacturing costs being as low as possible.

We have also explored the use of new polymer materials for polymer LEDs (PLEDs) with Ahmed Iraqi in the department of chemistry. He has designed and synthesized a new class of polymers called poly(alkylcarbazoles), which are blue emitting polymers that in principle have a very high hole mobility and increased photo-stability.

Prototype Devices

Above left: Electroluminescence from a prototype laser-patterned polymer LED array of red, green and blue LEDs. Each LED is 1 mm^2. This work was done in collaboration with the UK companies MicroEmissive Displays Ltd. and Exitech Ltd. and was funded by EPSRC/DTI.

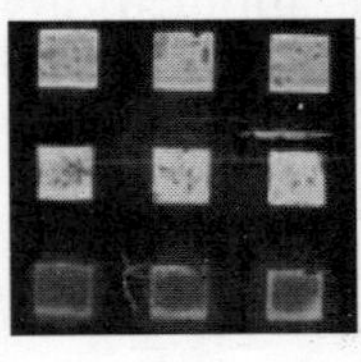

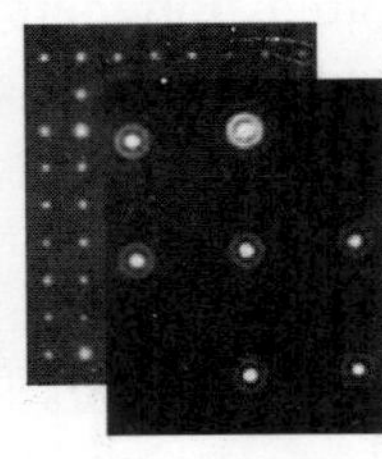

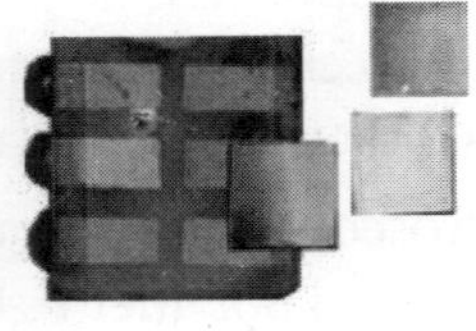

Above centre: Arrays of 200 (left) and 100 nm (right) diameter polymer LEDs created by the group. Nanoscale optical devices may be useful in a range of applications, including secure optical communications.

Above right: An interference lithographic technique developed by the group allows us to construct polymer LEDs on photonic crystals. Lithographically patterned LED substrates are shown inset top right, along with a working photonic device.

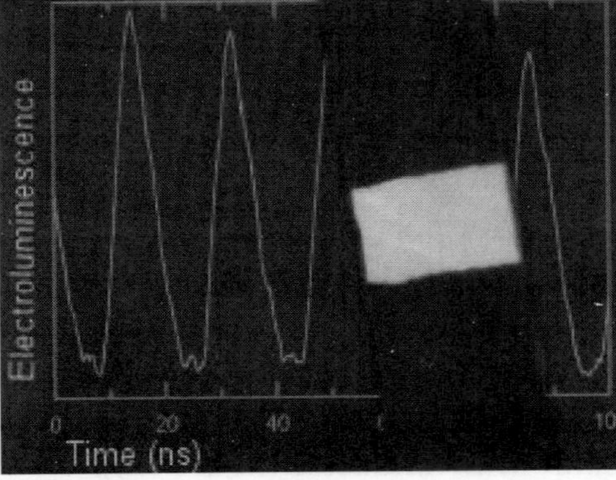

Right: A miniature (100 micron) LED specially designed by Iain Barlow, a member of the group, as a light-source for low-cost local-area optical communications applications. This device can generate well resolved electroluminescent pulses emitted at 56 MHz with a driving voltage of 3.8V (see graph in the background of the device). This technology could be used as the basis of an 'all plastic' optical communications system.

9.3 Electronic and Photonic Molecular Materials Group

The Electronic and Photonic Molecular Materials Group is part of the Molecular and Macromolecular Materials Group which has a dynamic research programme in the area of soft condensed matter.

The field of research into organic semiconductors has generated considerable interest due to the large number of novel properties found in this class of materials. It is now recognized as one of the most promising new areas of research in solid state physics.

Our research programme ranges from fundamental studies of the electronic excitations supported by conjugated polymers to the applications of such materials in technologies such as displays and electronics.

Areas of particular interest include:

- Spectroscopy of conjugated polymers
- Conjugated polymer light emitting devices
- Conjugated polymer field effect transistors
- Organic photonic devices and structures
- Micro and nanoscale characterisation of molecular materials

9.4 Organic Microcavities and Photonics

We have a long-standing interest in creating organic microcavities for both applications in displays and in the study of fundamental optical processes. A microcavity is a structure in which two high reflectivity mirrors are placed in close proximity—usually within half a wavelength (~ 200 to 400 nm). The mirrors quantize the optical field within the cavity, meaning that only photons of certain energy can be confined within the structure. Within the so-called 'strong-coupling' regime, the trapped cavity photons and the electronic states of a material placed in the cavity can undergo a mixing process, where the new states formed (termed cavity polaritons) are a superposition of optical and electronic states. We were the first group to demonstrate this effect in microcavities containing organic semiconductors (see figure 9.1). We have since shown that cavity polaritons can be organic cavities based on two metallic mirrors (figure 9.2), wedged organic cavities (figure 9.3), and cavities containing an organic and inorganic semiconductor (figure 9.4).

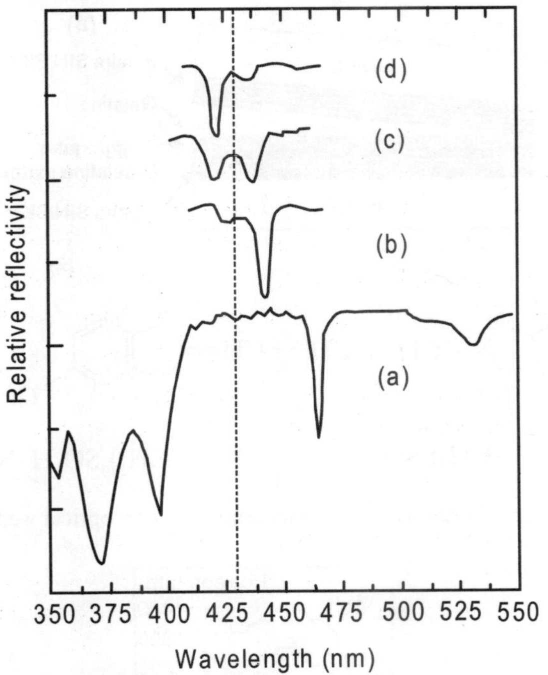

Fig. 9.1: Angle-tuned strong coupling in a organic microcavity.

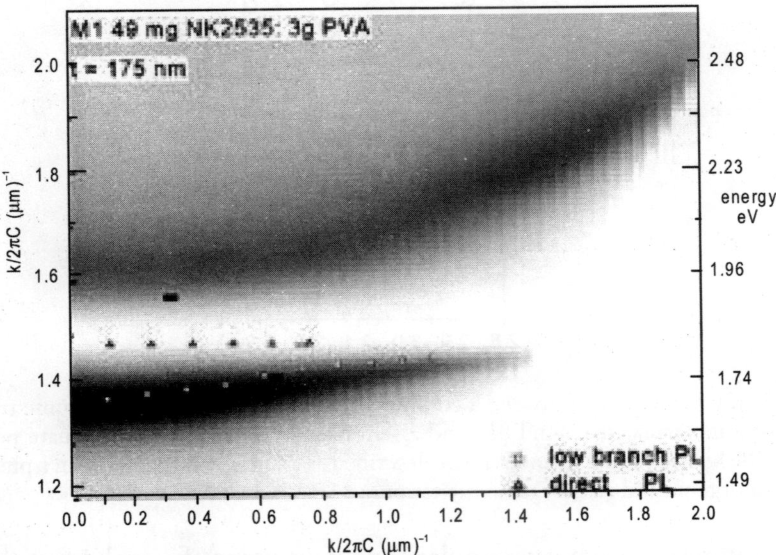

Fig. 9.2: Angle-tuned strong coupling in an all-metal organic microcavity.

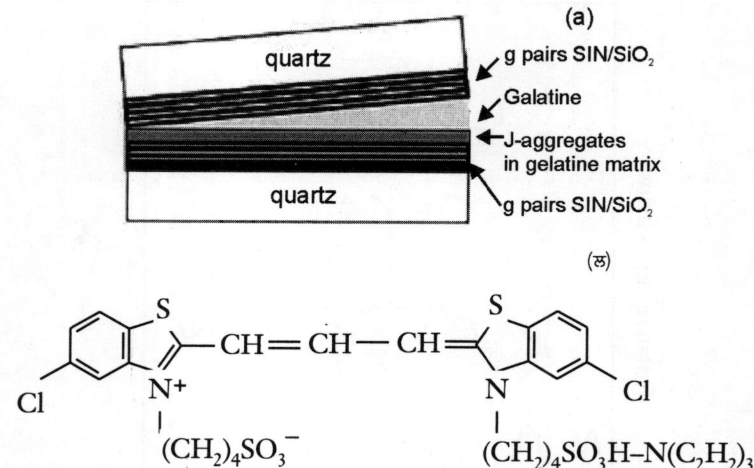

Fig. 9.3: A strong-coupled organic microcavity containing an optical wedge.

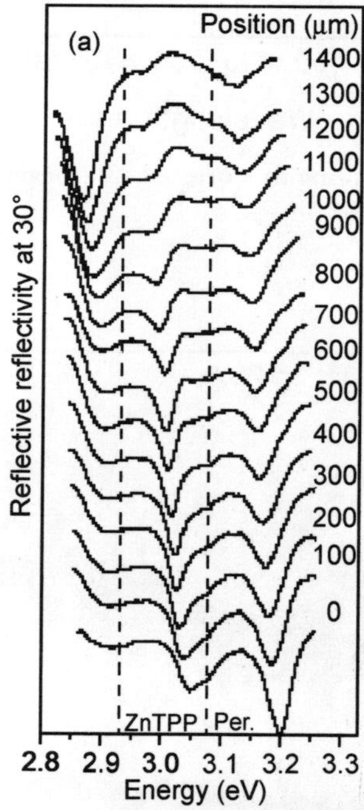

Fig. 9.4: 3-way anticrossing between a confined cavity-photon and the electronic transitions of a molecular dye (ZnTPP) and a self-assembled tetrabromoplumbate perovskite. The hybrid modes formed can be described as a superposition between a photon with an organic and an inorganic excitation.

The interest in such cavity polariton states comes from the fact that they have very different properties from their constituent components (cavity photons and electronic excitations [excitons]). This opens up many interesting applications, ranging from threshold-less lasers to ultra-fast optical switches. The advantage of using organic materials in the place of inorganics results from the fact that 'strong-coupling' effects can be easily observed at room temperature, and the high oscillator strength of organic excitations results in enhanced interaction with cavity photons. This enhanced interaction between light and matter when using organic materials to reach the strong-coupling regime means that effects can be realized in relatively low finesse optical structures.

9.5 Scanning Probe Microscopy

Scanning near field optical microscopy (SNOM) is a scanning probe technique that permits to diffraction limit to be broken by making optical measurements in the near-field of a surface. This is illustrated below left, where we overlay the photo luminescence intensity from a phase-separated polymer blend (F8 and PMMA) onto the surface topography. This nicely illustrates that SNOM can be used to gain both spectroscopic and topographic information simultaneously.

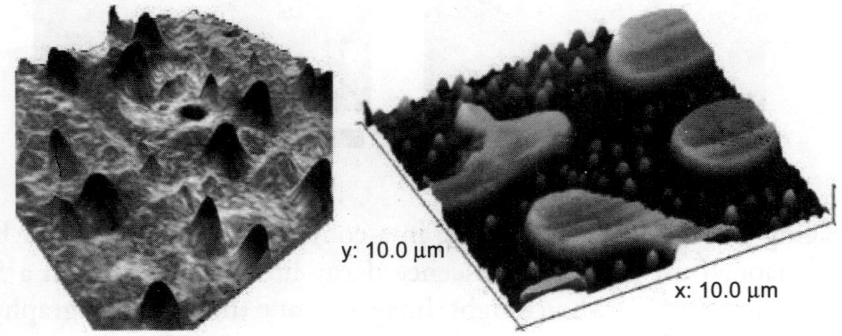

Above left: Fluorescence emission intensity superimposed on topography from an F8: PMMA blend film as measured by SNOM.

Above right: Fluorescence emission lifetime (shown using lighter colours) superimposed on topography from an F8: PMMA blend film as measured by SNOM.

We have developed our SNOM to perform time-resolved measurements, permitting a surface to be mapped with a spatial resolution of 70 nm and a temporal resolution of 60 ps. This technique has been used to study a number of systems, including the phase-separated F8: PS polymer blend shown in fig 9.2. Here, it can be seen that the fluorescence lifetime of the F8 varies significantly across the film, being significantly longer when it is trapped within the PS phase. We believe that this effect results from suppressed exciton diffusion to non-radiative defects.

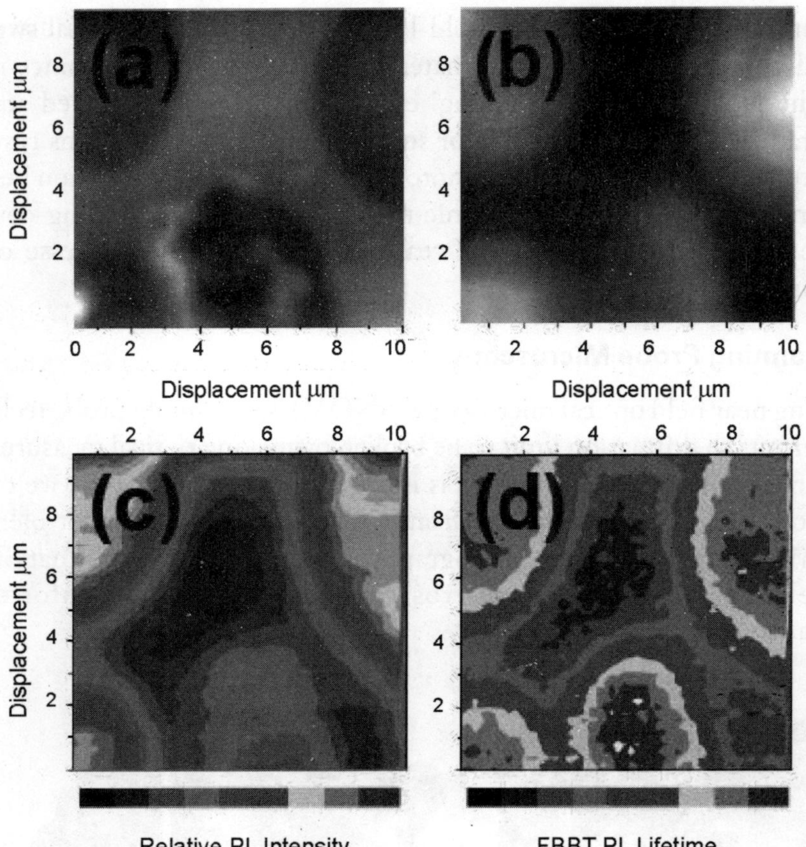

Imaging the fluorescence lifetime in a conjugated-polymer blend. Here, we have mapped the photoluminescence decay lifetime of F8BT in a 50:50 blend of F8 and F8BT, as shown right. Images (a) and (b) show topography and c.w. F8BT photoluminescence intensity respectively. The time-resolved intensity (at t=0) and decay lifetime are shown in (c) and (d). We find that the F8BT lifetime is almost twice as long when it is located in an F8-rich phase compared to a F8BT-rich phase. We believe that this results from suppression of excitons to non-radiative defects.

Handbook of Advanced Electronic and Photonic Materials and Devices

Overview of the Handbook

The Handbook of Advanced Electronic and Photonic Materials and Devices consolidates up-to-date research developments in semiconductors, superconductors, ferroelectrics, silicon-based materials, glasses, liquid crystals, light-emitting materials, conducting polymers, sol-gel materials, nonlinear optical and optoelectronic materials, electrochromic materials, laser materials, photoconductors, photovoltaic materials, low-k dielectrics, nanostructured materials, supramolecular and self-asemblies, functional polymers, photosynthetic and respiratory proteins, etc. The device applications of these materials have also been discussed in current semiconductor and photonic industries, computers, internet, information processing and storage, telecommunications, satellite communications, integrated circuits, optical fibers, microlasers, photocopiers, solar cells, batteries, light-emitting diodes, liquid crystal displays, magneto-optic memories, audio and video systems, recordable compact discs, video cameras, X-ray technology, color imaging, printing, flat-panel displays, optical waveguides, cable televisions, computer chips, nanotechnology, molecular-sized transistors and switches, as well as other emerging cutting edge technologies.

The Handbook is a multidisciplinary reference source for scientists and students working in the field of materials science, solid-state physics, electrical and optical engineering, chemistry, polymer science, ceramic and aerospace engineering, device and computational engineering, photophysics, biotechnology, data storage and information technology.

9.6 Electronic Materials and Devices

Much of the EMD Group research is driven by the following fundamental technological questions:

- How can the spectacular performance of Si integrated circuits be sustained as fundamental physical limitations arise?
- What alternative electronic device technologies can achieve performance specifications which exceed the capabilities of Si?

Since the gate length of today's silicon MOS devices are as small as 150 nm, addressing these issues necessarily requires expertise in all aspects of Nanotechnology—from nanofabrication using photon, ion and electron beams to self-assembly of organic molecular monolayers, to the coherent quantum properties of nanoclusters. The EMD plays a key role in the London Centre for Nanotechnology, a joint UCL—Imperial College collaboration with over £19M of infrastructure funding. This has led to collaboration with scientists from an increasingly diverse range of disciplines, including not only physicists and materials scientists, but also clinical medics and biopharmacologists.

EMD has a flourishing & diverse portfolio of areas technological excellence including:

- Ultra-Violet Processing of Ultrathin Films, in particular silicon dioxide and also high k layers, such as tantalum and hafnium oxides—materials for use as gate oxides in future Si MOSFET generations.

- Josephson Junctions: High-T superconductors for ultra-sensitive magnetic/electric field sensors
- Diamond Electronics, especially diamond-like carbon for use in high temperature environments and ultra-violet electronics applications where the higher bandgap of diamond is required.
- Dynamical Control Systems for process control in the biopharmaceuticals and other industries.
- Silicon Vacuum Electronics for sensor and field-emission display technologies.
- Si Nanoclusters: Exploitation of quantum effects for prototype quantum computers.

The group have been a central founding partner of the London Centre for Nanotechnology (LCN), are also very active in the Centre for Micro Biochemical Engineering and are co-investigating development programmes to fundamentally change many biopharmaceutical processes. A strong involvement exists with the joint Imperial College/UCL Centre for Process Systems Engineering in industrial monitoring and control.

Over the years, the group achieved top ratings in the RAE exercise. The average expenditure/member over the 6 years up to 2001 was £45k/annum, and the total number of publications was 318 (>50 per member—the highest in the department). Already, from 2001-2004, the total amount of the funding obtained already exceeds that for the previous 6 years. Future strategy includes, as follows:

- Development of new fully operational 220m2 clean room facility in the LCN.
- Taking a lead role in specification, installation, and management of Focused Ion Beam (FIB), Scanning Electron Microscope (SEM), Pulsed Laser Deposition (PLD) Plasma processing equipment, electrical characterisation, etc.
- Investigation of new materials and strategies for semiconducting, superconducting and magnetic devices, switches, Josephson junctions, and memories.
- Exploration of new growth systems for diamond, black diamond, and other carbon-based structures, including nanotubes, and polymers for a wide range of novel applications and heterostructure devices.
- Development of new deep ultraviolet light sources for applications into large area low temperature processing, of high k materials and other advanced oxides.

- ❖ Analysis of nonlinearities of plasma harmonics and sideband noise which are uniquely related to dynamic complex impedance and chemistry.
- ❖ Exploration of new control and monitoring applications, modeling predictive controllers, and root cause diagnosis and advanced cause and effect reasoning
- ❖ Study of single atom rare-earth doped materials and nanoclusters for telecomms (optical amplification, and photodetection) and quantum computing.
- ❖ Characterisation of light emission from dislocations in Si and SiGe: in particular, control using impurity ion emission
- ❖ Labelling of biological molecules, self-organisation of metallic and/or semiconductor nanowires.

10

Advanced Polymer Research

Polymers are long chain molecules with properties dominated by their chain behaviour and the nature of their chemical make-up or constitution. The distinction between thermoplastics and thermosets has become rather blurred with the development of new materials for more demanding environments than previously. They include high performance polymers which are more resistant to high temperatures, possess greater moduli or strengths, and can be combined with additives to enhance their intrinsic properties yet further. An understanding of the atomic and molecular construction of polymers provides an insight into how improved materials can be developed, in the subject of molecular engineering. It includes an understanding of both molecular configuration and conformation.

The manufacture of polymers from oil and gas feedstocks is dominated by a handful of processes operated in large scale petrochemical complexes. Intermediates and monomers are often interconvertible so that supply and demand can be matched. However, in the thermal cracking of naphtha the production of olefins varies within relatively narrow margins, so that excess quantities of propylene, for example, may be produced at medium and high cracking severity. Ethylene is the major petrochemical building block for vinyl chain growth polymers, and benzene and *para*-xylene are the main sources of aromatic structures within polymer repeat groups. Polymer consumption continues to grow and the range of both general-purpose materials and speciality polymers has widened considerably over the past two decades. Within each polymer type the range of grades available to the design engineer has also continued to widen, owing to the close control over molecular mass and structure that is available with the latest polymerization catalysts. Copolymerization, for example, is a major way in which structure can be tailored at a molecular level to produce the desired balance of physical properties in the end product. It allows both T_g and T_m to be modified to match the temperature scale of exposure of the end product, as well as affecting other properties like toughness and stiffness. A compromise between conflicting property demands often needs to be reached in the selection of the appropriate grade for a particular product. But there is an extra level of complexity in the effect of processing—it can affect molecular orientation and crystallisation and hence end properties. In addition, the constraints of processing can severely restrict the grades which can be processed effectively. So some compromise between design for function and design for manufacture must be planned in order to maximise and exploit the extra design freedom that polymers offer.

Do This

Now you have completed this unit, you might like to:

- ❖ Post a message to the unit forum.
- ❖ Review or add to your Learning Journal.
- ❖ Rate this unit.

Try This

You might also like to:

- ❖ Find out more about the related Open University course
- ❖ Book a FlashMeeting to talk live with other learners
- ❖ Create a Knowledge Map to summarise this topic.

10.1 Polymer Science and Engineering: An Introduction

Simply stated, polymer science is the science of large molecules. By virtue of their size alone, macromolecules have certain unique chemical and physical properties. It is also a relatively new academic discipline (at least in the U.S.) and one that is characterized by extraordinary breadth. It involves aspects of organic chemistry, physical chemistry, analytical chemistry, contemporary physics (particularly theories of the solid state and solutions), chemical and mechanical engineering and, most recently, electrical engineering. Moreover, there is a growing demand for what can be called engineering technologists, those skilled in the art of designing processes for producing specific products. Clearly, no one person has an in-depth knowledge of all these fields. Most polymer scientists and engineers have a broad overview of the subject that for those doing research is supplemented by a more detailed knowledge of a particular area. To give a flavor of what the field involves, particular areas are discussed in the following sections.

Polymer Synthesis

Most plastics people agree that it is unlikely that we will see any new thermoplastic take the world by storm (i.e. achieve levels of production comparable to polyethylene or polystyrene), but it should be kept in mind that similar things were being said round about 1950, just before high density polyethylene and isotactic polypropylene made their debut. This time they may be right, however, for two very good reasons. First, all the monomers that can be readily polymerized already have been; second, commercializing a new commodity plastic would probably cost well in excess of $1 billion. In any event, polymer chemists have better things to do than bash their heads against thermoplastic walls. The action these days is in specialty polymers or in finding new catalysts to make commodity plastics more cheaply or with precisely defined chain structures to give controlled properties. The types of specialty polymers that are important include those with stiff chains and strong intermolecular attractions to give thermal resistance and high strength, or chains with the types of delocalized structures that result in unusual electronic and optical properties. These materials are produced in smaller quantities than "bulk" or commodity plastics, but are [3]value added[2] products that command a much higher price. The synthesis of new polymer materials is a challenging area for anyone with an interest in organic chemistry.

Characterization

What a chemist thinks he or she has made is not always the stuff that is lying around the bottom of his or her test tube. Accordingly, there is an enormous field based on analysis or characterization. This is an intriguing and exciting area because of recent advances in instrumentation, particularly those interfaced with high-powered yet small and relatively cheap dedicated computers. These novel analytical techniques are not only useful in studying new materials, but answering questions that have intrigued polymer scientists for decades. For example, spectroscopic techniques are used to examine local chemical structures and interactions in polymer systems. Electron microscopy and the scattering of electromagnetic radiation are used to characterize overall structure; how components of a system separate into various types of phases; how chains fold into crystals; and determining the shape of an individual chain in a particular environment. Some techniques are so expensive that national facilities are required, e.g. synchrotron radiation and neutron scattering, but these are accessible to research scientists in the field. Polymer characterization is an exciting field for those interested in the relationship of molecular structure to macroscopic properties.

Polymer Physical Chemistry

Paul Flory was awarded the Nobel Prize in Chemistry for his work in this area and if you get into this subject you will come across his name a lot. This is a subject that demands a knowledge of theory and the ability to perform carefully controlled experiments, often using the types of instruments mentioned above. (The area of characterization and polymer physical chemistry overlap considerably and they are artificially separated here merely to illustrate the different types of things polymer scientists do). The simplest way to get a "feel" for this subject is get a copy of Flory's book "Principles of Polymer Chemistry", still a classic after more than forty years, and scan the chapters on the theories of rubber elasticity, solution thermodynamics, phase behavior etc. This subject remains intriguing, with all sorts of neat stuff in the areas of polymer blends or alloys, polymer liquid crystals, block copolymers, dendrimers, and so on.

Polymer Physics

Polymer physics and polymer physical chemistry are overlapping disciplines that are not, in many cases, easily delineated. Historically, however, it is possible to point to an enormous impact by theoretical physicists starting in the late 1960's and early 1970's. Until then, most theory was based on almost classical physical chemistry, but a number of leading physicists (notable de Gennes in France and Edwards in England) started to apply modern theories of statistical physics to the description of long chain molecules. The result has been a revolution in polymer theory, one that is not easily assimilated by traditional polymer scientists, that is still ongoing.

Polymer physics is not confined to theory, however. Experimental polymer physics continues to focus on areas such as chain conformation, viscoelastic and relaxation properties, phenomena at interfaces, the kinetics of phase changes, and electrical and piezoelectric properties.

Polymer Engineering and Technology

Last, but by no means least, there is a vast area that is involved with chemical engineering (the large scale chemical synthesis and processing of polymers) mechanical engineering (studies of strength, fatigue resistance, etc.) and engineering technology (designing a process to make a product) applied to polymer materials. These areas are often interactive. For example, there is enormous interest in producing ultra-high strength polymer fibers. It turns out that even "common or garden" polymers like polyethylene or polypropylene can be processed to give "high-tech" materials. The trick is to align the chains as perfectly as possible; not an easy task. Once made, of course, the mechanical properties of these materials have to be determined. This involves more than stretching a fiber until it breaks. The mechanical properties of polymers are complicated by all sorts of factors (defects, relaxation processes etc.) and the field is an intriguing combination of mechanical measurements, structural characterization and theory.

Undergraduate Polymer Science and Engineering Option

In recent years, economic changes and the increasing sophistication of polymer manufacturing technologies have created a need for skilled personnel in the polymer manufacturing industry who understand materials processes and can implement new technology quickly.

For these reasons, the Polymer Science and Engineering Program offers two "study tracks." The polymer properties track emphasizes the traditional materials-based aspects of polymers including their structure, properties, rheology, and processing. The polymer processing track emphasizes polymer processing in a unique hands-on, 9-credit course that is a mixture of lectures and lab work on production scale equipment.

10.2 Material Answers

Material Answers is a technical consulting firm specializing in material science and engineering.

Our projects encompass characterization, modeling, design, development, and optimization of all material classes, including polymers, ceramics, metals, and semiconductors.

Our services include:

- failure analysis
- accident investigation
- design for performance
- rheological testing and analysis
- process optimization
- scale-up
- design for target microstructure
- interfacial control
- processing-property relationships
- conceptual process design

10.3 Doctor of Philosophy Programme in Fiber and Polymer Science

This multidisciplinary program brings together the disciplines of mathematics, chemistry, physics, and engineering for the development of the independent scholars versed in the fields of polymer, fiber, and textile science. The program is coordinated by the College of Textiles and leads to the degree of Doctor of Philosophy. Students majoring in the physical sciences, engineering, mathematics, textiles, and having at least a 3.0 grade point average out of 4.0 in their undergraduate major and have a master's degree will normally qualify for admission. For exceptionally qualified students, the master's degree requirement may be waived and the student can be admitted directly into the Ph.D. program.

The polymer, fiber, and textile sciences are concerned with polymeric materials and fibers produced from them; textile assemblies in one, two, and three-dimensional forms; and the chemistry of dyeing, finishing, and other wet processes. This broad field of study permits a wide range of useful concentrations. The candidate is expected to concentrate in one area and to acquire a reasonable perspective in other relevant areas. Generally, a student specializes in the areas of (1) polymer chemistry and synthesis, (2) fiber and polymer physics and physical chemistry, (3) the production, processing and properties of fibrous materials, or (4) chemistry of dyes, finishes, and their processes. The student's research is usually based within one of these areas or another suitable one.

General Requirements

Research Facilities

The College of Textiles building on the Centennial Campus houses teaching and research facilities valued at over $50 million. The facility contains a complete Model Manufacturing Facility (MMF) which makes it possible for students to participate in research in all manufacturing processes from fiber extrusion to garment assembly. The major pieces of manufacturing equipment are connected with the college's distributed computing network which permits studies on Computer Integrated Manufacturing and Management Systems.

In addition to the MMF, the building contains research laboratories and pilot plant facilities for carrying out research in polymerization, dyeing, finishing, polymer evaluation, process evaluation, process control, color science, long and short-staple processing, fiber extrusion, knitting, weaving, tufting, production of nonwoven fabrics of all types, and apparel manufacturing. In addition, there are completely equipped laboratories for physical, mechanical, and dynamic testing, and for research in polymer, fiber, and textile structures. Robotics, Kawabata Hand Evaluation, Comfort, Thermal Protection, Electrotechnology, CAD, CAM, CAE, CIM, Management Systems, virtual reality, and computer visualization are other specialized labs available within the college.

The Burlington Textiles Library has a collection of over 27,000 bound volumes and is one of the world's leading information centers for textile research. Comprising 13,000 square feet, the facilities include a reserve reading room, group study and seminar rooms, individual study carrels, and a multimedia room. Special resources include an extensive collection of textile research and trade journals, abstracts and indexes, standards and test methods, conference proceedings, the Speizman Collection of hosiery samples, and the Harriss Collection of fabrics. A fully computerized online catalog and an extensive array of online databases are available to all patrons who wish to do searches within the library facilities. The catalog and selected databases can be accessed on the Internet and through the NCSU libraries page at http://www.lib.ncsu.edu/

The College of Textiles has one of the most advanced computing facilities found in any university in the world. The computing hardware consists of 500 Intel PCs, 15 UNIX workstations, and 5 Macintosh computers. Graduate students have access to 100 computers in general-purpose student computing labs, 40 computers in graduate offices, and computers in research laboratories. Furthermore, the university provides student-computing laboratories throughout the campus for student use. Hundreds of computer programs are available to graduate students and the college's high-speed connection to the Internet provides graduate students with access to a wealth of information and resources throughout the world. This includes connectivity to the super computer located in North Carolina Supercomputing Center in Research Triangle Park (RTP).

Excellent analytical facilities are also available. Specialized equipment includes thermal analysis facilities (DSC); X-ray diffraction facilities; chromatographic equipment; rheometers; color science and fiber physics instrumentation; physical testing, and mechanical testing; polymer and dye synthesis; optical, acoustic, interference, and scanning electron microscopy; IR and FTIR spectroscopy; and acoustical testing. Facilities for UV, gamma, and electron irradiation are available, as are machine and electronic shop facilities. In addition, extensive electron microscopy and NMR facilities are available in cooperating departments on campus.

Research Programs Industry, the State of North Carolina, and several agencies of the federal government support research in the College of Textiles at an annual level of over $12 million. Major, long-term research efforts include work on composites, the fundamentals of fiber formation, including very high-speed melt spinning and new methods of spinning fiber from cellulose solutions;

increasing the energy efficiency of dyeing, finishing and yarn and fabric production; introducing automation and new material handling techniques in apparel manufacturing; developing comfortable and physiologically acceptable fabrics for protection against toxic agents and fire; nonwoven fabric research; new yarn manufacturing techniques; and synthesis of new dyes for improved lightfastness. Extensive research is also carried out in such management related areas as industry modeling, studies of quick response in the soft goods pipeline, and modeling trends in the apparel market. The major centers and consortia involved are:

The National Textile Center—University Research Consortium is a federally funded research program involving faculty at NC State, Auburn University, Clemson University and Georgia Institute of Technology. It supports research activities in materials, design, manufacturing and systems all related to the fiber-textile apparel manufacturing complex.

The National Science Foundation, The State of North Carolina and several industry sponsors fund the Nonwovens Cooperative Research Center. The NCRC will fund both fundamental and applied research on nonwoven technologies, products and applications. The Center for Research on Textile Protection and Comfort (T-PACC) serves as the nucleus of activity for coordinating and managing diverse research projects in the area of combined textile comfort and protection. T-PACC facilities incorporate the capabilities to conduct basic and applied research on the properties of textile materials which relate to both wearer comfort and safety when exposed to hazardous and/or adverse environmental conditions. These facilities have become a unique resource for academic research. The object of the Manufacturing Technology for Apparel Automation project is to oversee the design and building of flexible, automated garment assembly devices to reduce the labor required in the production of apparel or military sewn products.

10.4 List of Important Publications

This is a list of important publications in physics, organized by field.

Some reasons why a particular publication might be regarded as important:

- *Topic creator*—A publication that created a new topic
- *Breakthrough*—A publication that changed scientific knowledge significantly
- *Introduction*—A publication that is a good introduction or survey of a topic
- *Influence*—A publication which has significantly influenced the world
- *Latest and greatest*—The current most advanced result in a topic

Optics

Book of Optics

❖ Ibn al-Haytham (Alhacen)

Description: The *Book of Optics* (Arabic: *Kitab al-Manazir*, Latin: *De Aspectibus*) is a seven volume treatise on optics and physics, written by the Iraqi Arab Muslim scientist Ibn al-Haytham (Latinized as *Alhacen* or *Alhazen* in Europe), and published in 1021, when he was under house arrest in Cairo, Egypt. The book had an important influence on the development of optics, and science in general, as it drastically transformed the understanding of light and vision, and introduced the experimental scientific method. As a result, Ibn al-Haytham has been described as the "father of optics", the "pioneer of the modern scientific method", and the "first scientist". The *Book of Optics* has been ranked alongside Isaac Newton's *Philosophiae Naturalis Principia Mathematica* as one of the most influential books ever written in the history of physics.

❖ Christıaan Huygens—A Treatise on Light

Huygens' treatise on light was not appreciated in its time until much later due to the mistaken zeal with which formerly everything that conflicted with the cherished ideas of Newton was denounced by his followers. Despite that, Huygens attained a remarkably clear understanding of the principles of wave-propagation; and his exposition of the subject marks an epoch in the treatment of Optical problems.

❖ Augustin-Jean Fresnel—Papers on optical phenomena

Work by Thomas Young and Fresnel provided a comprehensive picture of the propagation of light.

The Frequency Comb Papers

❖ " Accurate Measurement of Large Optical Frequency Differences with a Mode-Locked Laser" Th.Udem, J. Reichert, R. Holzwarth, and T.W. H"ansch, Opt. Lett., 24, 881 (1999).

❖ "Measuring the Frequency of Light with Mode-Locked Lasers" J. Reichert, R. Holzwarth, Th.Udem, and T.W.H"ansch, Opt. Commun. 172, 59 (1999).

❖ "Optical Frequency Metrology", Th.Udem, R. Holzwarth, and T. W. H"ansch, Nature 416, 233 (2002).

Description: The Frequency comb technique was presented in few papers. The earlier presented the main idea but last is the one often cited. Importance:

Classical Mechanics

Philosophiae Naturalis Principia Mathematica

❖ Isaac Newton

Description: The *Philosophiae Naturalis Principia Mathematica* (Latin: "mathematical principles of natural philosophy", often *Principia* or *Principia Mathematica* for short) is a three-volume work by Isaac Newton published on July 5, 1687. One of the most influential scientific books ever published, it contains the statement of Newton's laws of motion forming the foundation of classical mechanics as well as his law of universal gravitation. He derives Kepler's laws for the motion of the planets (which were first obtained empirically).

In formulating his physical theories, Newton had developed a field of mathematics known as calculus.

Importance: Topic creator, Breakthrough, Influence

Méchanique Analytique

❖ Joseph Louis Lagrange (1788)

Description: Lagrange's masterpiece on mechanics and hydrodynamics. Based largely on the calculus of variations, this work introduced Lagrangian mechanics including the notion of virtual work, generalized coordinates, and the Lagrangian. Lagrange also further developed the principle of least action and introduced the Lagrangian reference frame for fluid flow.

Classical Mechanics

❖ Herbert Goldstein

Description: The standard graduate textbook on classical mechanics, widely considered to be one of the best books on the subject.

Importance: Introduction, Influence

Special Theory of Relativity

On the Electrodynamics of Moving Bodies

❖ Albert Einstein
❖ Annalen der Physik. June 30, 1905
❖ On the Electrodynamics of Moving Bodies, Zur Elektrodynamik bewegter Körper (German original)

Description: Special relativity, developed in 1905, only considers observers in inertial reference frames which are in uniform motion with respect to each other. Einstein's paper that year was called "*On the Electrodynamics of Moving Bodies*". While developing this theory, Einstein wrote to Mileva (his wife) about "our work on relative motion". This paper introduced the special theory of relativity, a theory of time, distance, mass and energy. The theory postulates that the speed of light in vacuum will be the same for these observers. Special relativity solved the puzzle that had been apparent since the Michelson-Morley experiment, which had failed to show that light waves were travelling through any medium (other known waves travelled through media—such as water or air). It had been suggested that light waves actually did not travel through any medium: the speed of light was thus fixed, and not relative to the movement of the observer. This was impossible under Newtonian classical mechanics however, and Einstein provided a new system which allowed for this.

Importance: Topic creator, Breakthrough, Influence

The Theory of Relativity

- Ludwik Silberstein
- Cambridge University Press, 1914

Description: This pioneering textbook drew together the now well-known developments of H. A. Lorentz, A. Einstein, and H. Minkowski. It uses concepts developed in the then- current textbooks (e.g. Vector Analysis (Gibbs/Wilson) and Bonola: Non-Euclidean Geometry) to provide entry into mathematical physics including a vector-based introduction to quaternions and a primer on matrix notation for linear transformations of 4-vectors. The ten chapters are composed of 4 on kinematics, 3 on quaternion methods, and 3 on electromagnetism. The second edition published in 1924 extended relativity into gravitation theory with tensor methods, but was superseded by Eddington's text. The book has a conversational style and embellished with appropriate footnotes. While the mathematics is generally well adapted to the text, there is an erroneous expression given for the quaternionic representation of Lorentz transformations. The expression should have the form of an inner automorphism but Silberstein inexplicably uses the expression Q[]Q, failing to supply one of the Q's with a - 1 exponent. The actual technique in geometric arithmetic comes about with inversive ring geometry applied to biquaternions.

Importance: Influence

[Spacetime Physics

- Edwin F. Taylor, John Archibald Wheeler
- W. H. Freeman (2nd edition 1992) ISBN 0716723271

Description: A modern introduction to special relativity, that explains well how the choice to divide spacetime into a time part and a space part is no different than two choices bout how to assign coordinates to the surface of the earth. Suitable for self-study.

Importance: Introduction

General theory of Relativity

The Foundation of the General Theory of Relativity

- Albert Einstein
- Annalen der Physik., 1916
- The Foundation of the General Theory of Relativity, Die Grundlage der allgemeinen Relativitätstheorie (German original)

  Importance: Topic creator, Breakthrough, Influence

The Mathematical Theory of Relativity

- Arthur Stanley Eddington
- Cambridge University Press, 1923, 1924

Description: This textbook is a tour-de-force of tensor calculus, developed in Chapter II. By page 83 he has deduced the Schwarzschild metric for the domain of events around an isolated massive particle. By page 92 he has explained the advance of the perihelion of the planets, the deflection of light, and displacement of Fraunhofer lines. Electromagnetism is relegated to Chapter VI (pp. 170-195), and later (p. 223) The bifurcation of geometry and electrodynamics. This text, with its ambitious development of pseudo-Riemannian geometry for gravitational theory, set an austere standard with relativity enthusiasts. Gone is any mention of quaternions or hyperbolic geometry since tensor calculus subsumes them. Thus for learning the mechanics of modern relativity this text still serves, but for motivation and context of the special theory, Silberstein is better.

Importance: Influence

Gravitation

- Charles W. Misner, Kip S. Thorne, and John Archibald Wheeler
- W. H. Freeman, 1973

Description: A book on gravitation (often considered the "Bible" by researchers for its prominence) by Misner, Thorne, and Wheeler. Published by W. H. Freeman and Company in 1973. A massive tome of over 1200 pages, the book covers all aspects of the General Theory of Relativity and also considers some extensions and experimental confirmation. The book is divided into two "tracks", the second of which covers more advanced topics.

Importance: Introduction, Influence

A First Course in General Relativity

- Bernard F. Schutz
- Cambridge University Press, 1985

Description: A book on the Theory of General Relativity that is suitable for a year-long undergraduate course on the subject that can also stand as a semester-long course for graduate students.

Importance: Introduction

Quantum Theory

On the Law of Distribution of Energy in the Normal Spectrum

- Max Planck
- Annalen der Physik, vol. 4, p. 553 ff (1901).
- On the Law of Distribution of Energy in the Normal Spectrum

Description: In physics, the intensity spectrum of electromagnetic radiation from a black body at temperature T is given by the Planck's law of black body radiation:

$$I(\upsilon)d\upsilon d\Omega = \frac{2h\upsilon^3}{C^2}\frac{1}{\exp\left(\frac{h\upsilon}{kT}\right)-1}d\upsilon d\Omega$$

where:

v is the frequency

$I(v)$ is the amount of energy per unit time per unit surface per unit solid angle emitted in the frequency range between v and $v+\delta v$ [W m^{-2} Hz^{-1} sr^{-1}];

h is Planck's constant,:

c is the speed of light and

k is Boltzmann's constant.

Max Planck originally produced this law in 1900 (published in 1901) in an attempt to interpolate between the Rayleigh-Jeans law (which worked at long wavelengths) and Wien's law (which worked at short wavelengths). He found that the above function fit the data for all wavelengths remarkably well.

This paper is considered to be the beginning of quantum theory.

Importance: Topic creator, Breakthrough, Influence

The Principles of Quantum Mechanics

- P. A. M. Dirac

Description: Quantum mechanics as explained by one of the founders of the field, Paul Dirac. First edition published on 29 May 1930.

Importance: Introduction, Influence, Historical importance. The second to the last chapter is particularly interesting because of its prediction of the positron.

Table of Contents

1. The Principle of Superposition
2. Dynamical Variables and Observables
3. Representations
4. The Quantum Conditions
5. The Equations of Motion
6. Elementary Applications
7. Perturbation Theory
8. Collision Problems
9. Systems containing several similar particles
10. Theory of Radiation
11. Relativistic Theory of the electron
12. Quantum Electrodynamics

Principles of Quantum Mechanics

- R. Shankar
- Plenum Press (2nd edition 1994) ISBN 0-306-44790-8

Description: Standard graduate textbook in quantum mechanics.
Importance: Introduction

Introduction to Quantum Mechanics

- David J. Griffiths
- Prentice Hall (2nd edition 2004) ISBN 0-13-111892-7

Description: A how-to for Quantum Mechanics aimed at the physics undergraduate.
Importance: Introduction

Thermodynamics

An Experimental Enquiry Concerning the Source of the Heat which is Excited by Friction

- Benjamin Thompson, Count Rumford (1798). "An Experimental Enquiry

Concerning the Source of the Heat which is Excited by Friction". *Philosophical Transaction of the Royal Society*: 102.

Description: Observations of the generation of heat during the boring of cannons led Rumford to reject the caloric theory and to contend that heat was a form of motion.

Importance: Influence

On the Equilibrium of Heterogeneous Substances

- Gibbs, J. Willard (1875-1878). *On the Equilibrium of Heterogeneous Substances.* Connecticut Acad. Sci. Reprinted in
  - Gibbs, J. Willard (October 1993). *The Scientific Papers of J. Willard Gibbs (Vol. 1)*. Ox Bow Press. ISBN 0-918024-77-3.
  - Gibbs, J. Willard (February 1994). *The Scientific Papers of J. Willard Gibbs (Vol. 2)*. Ox Bow Press. ISBN 1-881987-06-X.

Description: Between 1876 and 1878 Gibbs wrote a series of papers collectively entitled "*On the Equilibrium of Heterogeneous Substances*", considered one of the greatest achievements in physical science in the 19th century and the foundation of the science of physical chemistry. In these papers Gibbs applied thermodynamics to the interpretation of physicochemical phenomena and showed the explanation and interrelationship of what had been known only as isolated, inexplicable facts. Gibbs' papers on heterogeneous equilibria included:

- Some chemical potential concepts
- Some free energy concepts
- A Gibbsian ensemble ideal (basis of the statistical mechanics field)
- A *phase rule*

Importance:

Classical and Statistical Thermodynamics

- Ashley H. Carter
- Benjamin Cummings (2000) ISBN 0137792085

Description: Covers the phenomenological basis of classical thermodynamics itself and also the statistical theory, without assuming the reader already knows statistics or quantum mechanics. Truly an introductory text, you can pick it up after taking advanced calculus and first year general physics and a semester later know about Bose-Einstein condensation, population inversions, and even information theory.

Importance: Introduction

Statistical Mechanics

On the Motion—Required by the Molecular Kinetic Theory of Heat—of Small Particles Suspended in a Stationary Liquid

- Einstein, Albert (1905). "Über die von der molekularkinetischen Theorie der Wärme geforderte Bewegung von in ruhenden Flüssigkeiten suspendierten Teilchen (On the Motion—Required by the Molecular Kinetic Theory of Heat—of Small Particles Suspended in a Stationary Liquid)". *Ann. Phys.* 17 (549).

Description: In this publication Einstein covered his study of Brownian motion, and provided empirical evidence for the existence of atoms.

Importance:

Scaling Laws for Ising Models Near T_c

- Leo P. Kadanoff
- Physica 2, p. 263 (1966).

Description: Introduces the real space view on the renormalization group, and explains using this concept some relations between the scaling exponents of the Ising model.

Importance: Topic creator, breakthrough, influence

The Renormalization Group: Critical Phenomena and the Kondo Problem

- Kenneth Wilson
- Rev. Mod. Phys. 47, 4, p. 773-840 (1974)

Description: Application of the renormalization group to the solution of the Kondo problem. The author was awarded the Nobel Prize in 1982 because of this work.

Importance: Breakthrough, influence

Equation of State Calculations by Fast Computing Machines

- Nicholas Metropolis, Arianna W. Rosenbluth, Marshall N. Rosenbluth, Augusta H. Teller, and Edward Teller
- N. Metropolis, A. W. Rosenbluth, M. N. Rosenbluth, A. H. Teller, and E. Teller (1953). "Equation of State Calculations by Fast Computing Machines". *Journal of Chemical Physics* 21 (6): 1087–1092. doi: 10.1063/ 1.1699114.

Description: Introduces the Metropolis Monte Carlo method with periodic boundary conditions and applies it to the numerical simulation of a fluid.

Importance: Topic creator

Electromagnetism

Experimental Researches in Electricity

- Michael Faraday
- Experimental Researches in Electricity, vols. i. and ii., Richard and John Edward Taylor, vols. i. and ii. (1844 and 1847); vol. iii. (1844); vol. iii. Richard Taylor and William Francis (1855);
- "Experimental Researches in Electricity" by Michael Faraday Original text with Biographical Introduction by Professor John Tyndall, 1914, Everyman edition.

Description: Faraday's law of induction and research in electromagnetism

Importance:

A Dynamical Theory of the Electromagnetic Field

- James Clerk Maxwell
- Maxwell, James Clerk, "A Dynamical Theory of the Electromagnetic Field". 1865.

Description: "A Dynamical Theory of the Electromagnetic Field" was the third of James Clerk Maxwell's papers concerned with electromagnetism. The concept of displacement current was introduced, so that it became possible to derive equations of electromagnetic wave. It was the first paper in which Maxwell's equations appeared.

Importance: Topic creator, breakthrough, influence

Classical Electrodynamics

- John David Jackson
- Wiley (3rd edition 1998) ISBN 047130932X

Description: The defining graduate-level introductory text.

Importance: Influence, Introduction

Introduction to Electrodynamics

- David Griffiths
- Prentice Hall. (3rd edition 1998) ISBN 0-13-805326-X.

Description: A standard undergraduate introductory text.

Importance: Introduction

Fluid Dynamics

An Experimental Investigation of the Circumstances which Determine whether the Motion of Water shall be Direct or Sinuous, and of the Law of Resistance in Parallel Channels

- Osbourne Reynolds
- Philosophical Transactions, vol. 174, (1883).

Description: Introduces the dimensionless Reynolds number, investigating the critical Reynolds number for transition from laminar to turbulent flow.

The Local Structure of Turbulence in Incompressible Viscous Fluid for Very Large Reynolds Numbers

- A. N. Kolmogorov
- Dokl. Akad. Nauk. SSSR 30, p. 4 (1941). Reprinted in Proc. Roy. Soc. A 434, p. 9 (1991).

Description: Introduces the only quantitative theory on turbulence which has survived the test of time.

Importance: Breakthrough, Influence

Statistical Fluid Mechanics

- A.S. Monin, A.M. Yaglom
- The MIT press (1971). First edition in Russian by Nauka (1965).

Description: The most important review text on turbulence.

Importance: Relevant textbook.

Nonlinear Dynamics and Chaos

Deterministic Nonperiodic Flow

- Edward Lorenz
- Journal of Atmospheric Sciences, vol. 20, p. 130-148 (1963).

Description: A finite system of deterministic nonlinear ordinary differential equations is introduced to represent forced dissipative hydrodynamic flow, simulating simple phenomena in the real atmosphere. All of the solutions are found to be unstable, and most of them nonperiodic, thus forcing to reevaluate the feasibility of long-term weather prediction. In this paper the Lorenz attractor is presented for the first time, and gave the first hint of what is now known as butterfly effect.

Importance: Topic creator, Breakthrough

Period Three Implies Chaos

- T. Y. Li, J. A. Yorke.
- American Mathematical Monthly 82, pp. 985-992, (1975).

Quantum Field Theory

Space-Time approach to Quantum Electrodynamics

- Richard P. Feynman
- Physical Review, vol. 76, 6, p. 769 (1949).

Description: Introduction of the Feynman diagrams approach to quantum electrodynamics.

Importance: Topic creator, Breakthrough, Influence

An Introduction to Quantum Field Theory

- Michael E. Peskin and Daniel V. Schroeder
- Addison-Wesley Advanced Book Program (1995).

Description: Standard graduate textbook in quantum field theory.

Importance: Introduction

Cosmology

The Early Universe

- E. W. Kolb, M. S. Turner
- Addison-Wesley, 1990.

Description: Reference textbook on cosmology, discussing both observational and theoretical issues.

Importance: Relevant textbook.

A Preliminary Measurement of the Cosmic Microwave Background Spectrum by the Cosmic Background Explorer (COBE) Satellite

- J. C. Mather, *et al.*
- ApJ, v. 354, p. L37, (1990)
- Online version

Description: This paper was part of the COBE project. The COBE satellite was developed by NASA's Goddard Space Flight Center to measure the diffuse infrared and microwave radiation from the early universe to the limits set by our astrophysical environment. It was launched November 18, 1989 and carried three instruments, a Far Infrared Absolute Spectrophotometer (FIRAS) to compare the spectrum of the cosmic microwave background radiation to a precise blackbody, a Differential Microwave Radiometer (DMR) to map the cosmic radiation and search for brightness variants, and a Diffuse Infrared Background Experiment (DIRBE) to search for the cosmic infrared background radiation produced by the first galaxies.

FIRAS—The cosmic microwave background (CMB) spectrum is that of a nearly perfect blackbody with a temperature of 2.725 +/- 0.002 K. This observation matches the predictions of the hot Big Bang theory extraordinarily well, and indicates that nearly all of the radiant energy of the Universe was released within the first year after the Big Bang. Initial results from FIRAS were presented in this paper. Final results from FIRAS were presented at: J.C. Mather, *et al.*, " Calibrator Design for the COBE Far-Infrared Absolute Spectrophotometer (FIRAS)", ApJ, v. 512, p. 511 (1999)

Importance:

Structure in the Cobe Differential Microwave Radiometer First-year Maps

- G.F. Smoot *et al.*
- ApJ, v. 396, p. L1 (1992)
- Online version

Description: This paper was part of the COBE project. The COBE satellite was developed by NASA's Goddard Space Flight Center to measure the diffuse infrared and microwave radiation from the early universe to the limits set by our astrophysical environment. It was launched November 18, 1989 and carried three instruments, a Far Infrared Absolute Spectrophotometer (FIRAS) to compare the spectrum of the cosmic microwave background radiation to a precise blackbody, a Differential Microwave Radiometer (DMR) to map the cosmic radiation and search for Brightness variants, and a Diffuse Infrared Background Experiment (DIRBE) to search for the cosmic infrared background radiation produced by the first galaxies.

DMR—The CMB was found to have intrinsic "anisotropy" for the first time, at a level of a part in 100,000. These tiny variations in the intensity of the CMB over the sky show how matter and energy was distributed when the Universe was still very young. Later, through a process still poorly understood, the early structures seen by DMR developed into galaxies, galaxy clusters, and the large scale structure that we see in the Universe today. Initial results from FIRAS were presented in this paper. Final results from FIRAS were presented at: C.L. Bennett, *et al.*, " Four-Year COBE DMR Cosmic Microwave Background Observations: Maps and Basic Results", ApJ, v. 464, p. L1 (1996)

Importance:

The COBE Diffuse Infrared Background Experiment Search for the Cosmic Infrared Background. I. Limits and Detections

- M. G. Hauser, *et al.*
- ApJ, v. 508, p. 25 (1998)
- Online version

Description: This paper was part of the COBE project. The COBE satellite was developed by NASA's Goddard Space Flight Center to measure the diffuse infrared and microwave radiation from the early universe to the limits set by our astrophysical environment. It was launched November 18, 1989 and carried three instruments, a Far Infrared Absolute Spectrophotometer (FIRAS) to compare the spectrum of the cosmic microwave background radiation to a precise blackbody, a Differential Microwave Radiometer (DMR) to map the cosmic radiation and search for brightness variants, and a Diffuse Infrared Background Experiment (DIRBE) to search for the cosmic infrared background radiation produced by the first galaxies.

DIRBE—Infrared absolute sky brightness maps in the wavelength range 1.25 to 240 micrometres were obtained to carry out a search for the cosmic infrared background (CIB). The CIB was originally detected in the two longest DIRBE wavelength bands, 140 and 240 micrometres, and in the short-wavelength end of the FIRAS spectrum. Subsequent analyses have yielded detections of the CIB in the near-infrared DIRBE sky maps. The CIB represents a "core sample" of the Universe; it contains the cumulative emissions of stars and galaxies dating back to the epoch when these objects first began to form.

Importance:

Condensed Matter Physics

Theory of Superconductivity

- J. Bardeen, L. N. Cooper, and J. R. Schrieffer
- Phys. Rev. 108 (5), 1175-1204 (1957).

Description: The BCS theory of usual (not high T_c) superconductivity, relating the interaction of electrons and the phonons of a lattice. The authors were awarded with the Nobel prize.

Importance: Breakthrough, Influence

Solid State Physics

- Neil W. Ashcroft, N. David Mermin
- Brooks Cole, 1976, ISBN 0030839939

Description: It is so old that it still calls condensed matter physics by the out of fashion name of solid state physics, but yet it is still a good introduction to the topic.

Importance: Introduction

Simulating Physics with Computers

Description: A digital computer as an efficient universal computing device; the simulation of quantum mechanics and the use of quantum computers

Importance: influence

Polymer Physics

Path Integrals in Quantum Mechanics, Statistics, Polymer Physics, and Financial Markets

- Kleinert, Hagen
- 4th edition, World Scientific (Singapore, 2004); Paperback ISBN 981-238-107-4 *(also available online: PDF-files)*

The Theory of Polymer Dynamics

- M. Doi, S. F. Edwards, (International Series of Monographs on Physics)

On Intramolecular Statistics, Particularly for Chain Molecules

- Eugene Guth and Herman Mark,
- *Monatschefte für Chemie*, 65, 93 (1934).

Description: This paper contains, among other contributions, the first theoretical description of statistical mechanics of polymers with application to viscosity and rubber elasticity, and an expression for the entropy gain during the coiling of linear flexible molecules.

Importance: Contains the foundation of the *kinetic theory of rubber elasticity.*

Elastic and Thermodynamic Properties of Rubberlike Materials: A Statistical Theory

- Eugene Guth and Hubert M. James
- *Industrial Engineering Chemistry*, 33, 624 (1941)

Description: This work was presented earlier by Guth at the American Chemical Society meeting of 1939. The article contains the first outline of the *network* theory of rubber elasticity. The resulting Guth-James equation of state is analogous to van der Waal's equation.

Importance: Pioneering contribution to polymer physics.

Theory of Elastic Properties of Rubber

- Eugene Guth and Hubert M. James
- *Journal of Chemical Physics*, 15, 2941 (1943).

Description: This article presents a more detailed version of the network theory of rubber elasticity. The paper used average forces to some extent instead of thermodynamical functions. In statistical thermodynamics, these two procedures are equivalent.

Importance: Pioneering contribution. After some controversy within the literature, the James-Guth network theory is now generally accepted for larger extensions. See, e.g., Paul Flory's comments in *Proc. Royal Soc. A.* 351, 351 (1976).

Plasma Physics

The Collected Works of Irving Langmuir (1961)

- Irving Langmuir
- Vol. 3: Thermonic Phenomenon: papers from 1916-1937
- Vol. 4: Electrical Discharges: papers from 1923-1931

These two volumes from Nobel Prize winning scientist Irving Langmuir, include his early published papers resulting from his experiments with ionized gases (i.e. plasma). The books summarise many of the basic properties of plasmas. Langmuir coined the word *plasma* in about 1928.

Importance: Influence

Cosmical Electrodynamics, 2nd ed. (1963)

- Hannes Alfvén & Carl-Gunne Fälthammar

Hannes Alfvén won the Nobel Prize for his development of magnetohydrodynamics (MHD) the science that models plasma as fluids. This book lays down the ground work, but also shows that MHD may be inadequate for low-density plasmas such as space plasmas.

Importance: Topic creator, Breakthrough, Influence

Vehicle Dynamics

The Automotive Chassis Engineering Principles

- J. Reimpell H. Stoll J. W. Betzler.
- ISBN 978-0-7680-0657-5

Description: Vehicle dynamics and chassis design from a production car perspective.

Importance: Latest and greatest.

Race Car Vehicle Dynamics

❖ William F. Milliken and Douglas L. Milliken.

Description: Vehicle dynamics and chassis design from a race car perspective.

Importance: Latest and greatest, also the standard reference for automotive suspension engineers.

Fundamentals of Vehicle Dynamics

❖ Thomas Gillespie.

Description: Mathematically oriented derivation of standard vehicle dynamics equations, and definitions of standard terms.

Importance: Introduction to modern vehicle dynamics theory.

Chassis Design—Principles and Analysis

❖ William F. Milliken and Douglas L. Milliken.

Description: Vehicle dynamics as developed by Maurice Olley from the 1930s onwards. First comprehensive analytical synthesis of vehicle dynamics.

Importance: Topic creator.

Tyre Modelling for Use in Vehicle Dynamics Studies

❖ Bakker, E.; Nyborg, L.; Hans B. Pacejka

Description: A new way of representing tyre data obtained from measurements in pure cornering and pure braking conditions.

Importance: A standard reference in vehicle dynamics.

Geophysics

Seismic Data Processing

❖ Ozdogan Yilmaz

❖ Society of Exploration Geophysicists, 1987, (ISBN 0-931830-40-0).

Description: Up to date account of seismic data processing in the petroleum geophysics industry.

Importance:

Mathematical Physics

Invariante Variationsprobleme

❖ Emmy Noether (1918)

Description: Contained a proof of Noether's Theorem (expressed as two theorems), showing that any symmetry of the Lagrangian corresponds to a conserved quantity. This result had a profound influence on 20th century theoretical physics.

Ising's Thesis

- Ising's 1924 thesis proving the non-existence of phase transitions in the 1-dimensional Ising model.

Contour Argument

- Peierls' 1936 contour argument proving the existence of phase transitions in higher dimensional Ising models.

Infrared Bounds, Phase Transitions and Continuous Symmetry Breaking

- Jürg Fröhlich, Tom Spencer, and Barry Simon
- Infrared bounds, phase transitions and continuous symmetry breaking

Description: Proved the existence of phase transitions of continuous symmetry models in at least 3 dimensions. Communications in Mathematical Physics 50 (1) p79-95 1976

Importance:

Mathematical Physics (1961)

- Donald H. Menzel, Harvard University
- Dover Publications, 1961

Description: Thorough introduction to the mathematical methods of classical mechanics, electromagnetic theory, quantum theory and general relativity. Possibly more accessible than Morse and Feshbach. First published 1961. Available in Dover Editions.

Importance: Introduction.

Physics of Computation

Lloyd, S., 2000, Ultimate physical limits of computation, *Nature*, 406 : 1047-1054.

10.5 Design and Manufacture with Polymers

The polymer industry changes rapidly with the continuing introduction of new materials and processes. This course will update your knowledge of polymer design both in processing and manufacturing, and for function and use. You will examine the chemical formulation and physical properties of polymers, and their implications for design. You will explore polymer processing and examine some of the environmental issues raised by polymer materials. You will also look at recent developments in moulding and examine examples of products and product failure. You will assess the cost-effectiveness of polymer processes, look at management tools for improving quality and productivity, and develop your awareness of good design practice for polymer products.

Course Content

The polymer industry is changing fast, with the introduction of new materials and new processes. This course will update your knowledge of polymers from design, through manufacturing processes, to product use. It will develop your understanding of polymer properties, your skills in assessing cost effectiveness of polymer processes, and your awareness of good design practice.

The course is arranged in six blocks that broadly cover two distinct areas: design for processing and manufacturing, and design for function and use.

Block 1 Introduction to polymers Structure of long-chain molecules: repeat block and molecular mass distribution. Chemical formation of polymers: industrial manufacture of monomers from natural gas and oil; commercial polymerisation to give synthetic resins. Physical properties of polymers: viscoelasticity, crystallisation and orientation. The engineering design of a small dinghy.

Block 2 Solid properties and design Effects of viscoelasticity on end-use property: creep, stress relaxation and rupture. The ultimate response of polymer products: fracture mechanics; implications for design. Thermal properties of polymers: importance for processing to shape and for service conditions. Electrical and optical properties of polymers with some practical examples. Examples of product failure.

Block 3 Polymer processing Shaping polymers and growth rate. Melt viscosity and rate of shear. Cycle times and heat flow during cooling. Additives that modify service properties. Recent developments in rotational moulding.

Block 4 Part 1 Extruded products Design and manufacture of gas pipes. Specification for the final pipe and the process of extrusion. Extruder output: flow, performance. Energy balance and heat transfer.

Block 4 Part 2 Injection moulding Optimising production in the main moulding method: polymer melt flow in narrow channels. Analysis of flow, polymer type and control of heat flow, and relation to design of the moulding machines and mould design. Using CAD methods and examples of recent products. Recent developments such as gas moulding. Management tools for improving product quality (such as FMEA, failure modes and effects analysis); ways of improving productivity, such as TQA (total quality assessment) and JIT (just-in-time) manufacture.

Block 5 Part 1 Fibres and fibre assemblies The main natural and synthetic organic fibres: relating macromolecular structure to physical properties. High-performance fibres: strength of covalent bonds in polymer chains. Biomolecular engineering: the importance for the synthesis of biomaterials using recombinant DNA technology. Textiles: their real or potential use in composites.

Block 5 Part 2 Polymer composites Mixing powdered or fibrous fillers into both thermoplastic and thermosetting polymer resins. Stress response of a fibrous composite: the fibre size and volume fraction. Types of fibre: glass, carbon and aramid. Manufacturing methods: vehicle bodies and various filament-wound products such as prop shafts. Natural materials such as wood.

Block 6 Elastomers and adhesives Modelling of conventional vulcanised rubber parts: stress response. Key rubber parts for engineering structures: seals, bushes, and springs for engines and bearings for bridges. Manufacture of key rubber parts: heat transfer and control of vulcanisation. Thermoplastic elastomers: the very wide range of physical properties; processing by thermoplastic injection moulding. Adhesives: extra freedom in assembly, effects on other areas of manufacture. Some environmental issues raised by polymer materials.

Entry

There are no entry requirements, but we do assume that you have already done some study, up to HNC, HND or bachelors degree level, in a relevant subject area, or have equivalent experience from your employment. The course is designed for technically qualified managers in the polymer industry, mechanical engineers, manufacturing engineers, engineering designers and teachers of materials technology and engineering. If you have any doubt about the suitability of this course, please contact our Student Registration & Enquiry Service.

Qualifications

T838 is a specified course in our

- Joint Postgraduate Diploma in Computing and Manufacturing
- Postgraduate Diploma in Manufacturing: Management and Technology
- MSc in Manufacturing: Management and Technology
- MSc in Science

It can also count towards our Postgraduate Diploma in Engineering (E22), MEng (M03), MA in Online and Distance Education (F10) and MSc in Engineering (F46). We advise you to refer to the relevant award descriptions for information on the circumstances in which the course can count towards these qualifications because from time to time the structure and requirements of a qualification may change.

Excluded Combinations

If this course is in a similar area to one you have already completed, you can find out if it is an excluded combination.

If You have a Disability or Additional Requirements

Some of the course materials are presented on video cassette, some on audio cassette and some on floppy disk. You will need to spend considerable amounts of time using a personal computer and the internet. If you are a new student, or new to courses using a computer or the internet, make sure that you have our booklet Meeting Your Needs. You can obtain a copy by contacting our Student Registration & Enquiry Service. We provide a range of support services for individual needs but some of these may take several months to arrange. The website www.open.ac.uk/disability has the latest information about availability. Please contact us for advice if you have concerns about taking this course, or the support that could be provided.

Course Materials

What's included

Course books, other printed materials, audio cassette, video cassettes, software, home kit.

You will Need

Video and audio cassette players. You will need internet access and a computer as described in our Personal Computing for OU Study section.

Teaching and Assessment

Support from your tutor

You will have a tutor who will help you with the course material and mark and comment on your written work, and whom you can ask for advice and guidance. You will be able to contact your tutor by telephone, email and post. An optional introductory Saturday tutorial may be offered at up to four centres (London, Walsall, Manchester and Glasgow) shortly after the beginning of the course, at no extra cost, subject to viable group sizes in all locations. There is usually a lively student online forum. Contact the Postgraduate Technology & Computing (PTC) Office at The OU in the East Midlands (telephone 0115 971 5566, email R05-postgrad@open.ac.uk) if you want to know more about study with The Open University before you register.

We offer a revision school at the end of the course to help you to prepare for the examination. It is an ideal opportunity to consolidate your learning, meet tutors and other students, clear up any problems and help your revision. The fee for the school (around £230) is not included in the course fee and places are limited. The school is organised by our PTC Office, who will send you details of how to apply for a place.

Assessment

Three tutor-marked assignments, which we would normally expect you to submit online using our eTMA system, and an examination.

Professional Recognition

This course is a core module in the joint MSc run with the University of North London under the auspices of the Integrated Graduate Development Scheme. The Open University is registered with The Institute of Quality Assurance (IQA), Institute of Materials (IOM) and The Institution of Occupational Safety and Health (IOSH), for professional recognition.

Course Starting Dates

The details given here are for the final course starting date in November 2008.

Related Websites

You may be interested in free and open educational resources from this course available on The Open University's openlearn website.

Students also Studied

Students who studied this course have also studied at some time:

- Personal and career development in engineering (T191)
- Towards chartership: professional development for engineers (T398)
- Forensic engineering (T839)
- MSc research course (T802)
- Manufacture materials design (T881)

If you are studying towards a qualification please read its description to help you decide which route through our courses and which level of study are most appropriate for you.

Registration Information

The fees include all the course material, study support and assessment. Any additional items that you will need (for example a personal computer) are listed in the 'You will need' section above. Other costs that you may need to plan for could include stationery, perhaps a dictionary, and travel to attend tutorials or residential schools if relevant to your course.

The fees can be paid by instalments and there are various forms of help available in the UK only, for those who would otherwise find it difficult to pay their fees.

Even if you are working and have a household income as high as £30,000 (more if you have dependants) you could be entitled to some financial support. Depending on where you live in the UK this may include financial awards to help pay your fees and a contribution towards study costs. See Financial support for more information.

Where this Course is Available for Study

This course is available for study in the United Kingdom, the Republic of Ireland and the following European OU study areas:

Andorra, Austria, Belgium, Bulgaria, Cyprus, Czech Republic, Denmark, Estonia, Finland, France, Germany, Gibraltar, Greece, Hungary, Italy, Latvia, Liechtenstein, Lithuania, Luxembourg, Malta, Monaco, Netherlands, Poland, Portugal, Romania, San Marino, Slovakia, Slovenia, Spain, Sweden, Switzerland and Vatican City State.

All teaching is in English and your proficiency in the English language should be adequate for the level of study you wish to take.

Most students outside the UK and those resident in the Channel Islands and Isle of Man, will have to pay a higher fee because the UK fee includes a government subsidy restricted to UK residents and students with British Forces Post Office addresses.

Find out about other courses available in your country.

How to Register

To register a place on this course return to the top of the page and use the *Click to register* button.

More information

For more information and advice about registration go to Help with Registration.

Events Near You

Why not come and meet us at one of our events where you can talk to Open University staff. At some events there may also be samples of course materials and information and advice about specific subject areas and careers.

10.6 Research Opportunities

Materials Research Opportunities are available to undergraduates at all levels. You may participate in research for credit, research for pay, or a combination of both. A partial listing of research opportunities available as of xx/xx/xx is shown below. We also encourage you to contact your favorite professors to find out about the most recent opportunities in their groups.

Materials Research Opportunities for Undergraduates by Course #/Class Standing/etc.

Program	*1st Year*	*2nd Year*	*3rd Year*	*4th Year*	*Credit*	*Pay*
MSE 280	×	×			×	
MSE 490			×	×	×	
REU	×	×	×	×		×
MSPS			×	×	×	×

REU: Research Experiences for Undergraduates
MSPS: Marian Sarah Parker Scholars Program

Materials Research Opportunities for Undergraduates by Materials Class

Materials Class	*MSE 280*	*MSE 490*	*REU*	*MSPS*
Biomaterials	×	×	×	×
Ceramics	×	×	×	×
Electronic Materials	×	×	×	×
Metals	×	×	×	×
Polymers	×	×	×	×

11
Polymer Fibre Applications

11.1 "Conducting" Polymer Composites: The Nanotechnology Way

Conventional Techniques

Metals: Metals like copper, silver and aluminium have always been the forerunner for conductivity applications. Subsequently, when otherwise insulating polymer matrices were attempted to be tuned for similar applications by addition of suitable filler materials, the obvious choices were once again similar metals, in the forms of powders, flakes and fibers. Stainless steel powder filled acrylonitrile-butadiene-styrene (ABS) was tried in power measurement recorders. Metallic silver filled elastomeric gaskets/adhesives and metal mesh embedded fiber composites are also used in electronic applications especially for EMI shielding requirements. It was observed that a minimum volume fraction (percolation threshold) of the conducting reinforcement was required to establish a continuous conductive network within the polymer matrix to affect a sudden drop in its bulk resistivity. It has been reported that addition of 19 volume % of silver powder makes polymer composites electrically conductive. However, this comes with a heavy penalty on the weight, mechanical behaviour and cost.

Carbon: As a result, the search for better alternatives continued. The next obvious choice was carbon. Carbon had been long used as conductive fillers in polymer matrices, both in crystalline (graphite) and amorphous forms (carbon black). Electrically/thermally conducting carbon fiber, either in the form of chopped strands or as fabrics has been used to enhance the structural as well as the conductivity properties of the composite materials. These fibers are generally developed from Pitch or PAN based precursors, the choice of which plays an important role in the final properties of the fibers. This apart, other factors like the volume fraction and inherent conductivity of the fibers, the aspect ratio and their orientation are among the ones that play major roles in determining the conductivity properties of such composites.

Nanotube Reinforcements

Aspect Ratio: Other factors remaining constant, the composite conductivity for a fiber-reinforced matrix was reported to be directly proportional to the aspect ratio of the fibers6. The aspect ratio of the common carbon/graphite fibers is in the range of 10 to 50. However, for **carbonnanotubes** (CNT), which are essentially rolled up graphene sheet(s) having diameters in the range of 10 nm and length up to a few microns, the aspect ratio is often in the order of 500 and above.

Fiber Conductivity: Another important factor, which determines the composite conductivity, is the inherent conductivity of the fibers. In this regard, the CNTs offer some unique advantages over the conventional metallic or graphitic fibers. To appreciate that, one needs to have a basic understanding about the formation and structure of the CNTs.

CNT Structure: Single wall nanotubes (SWNT) are formed by rolling individual graphene sheets (one layer of crystalline hexagonal graphite) as shown in Fig. 11.1.

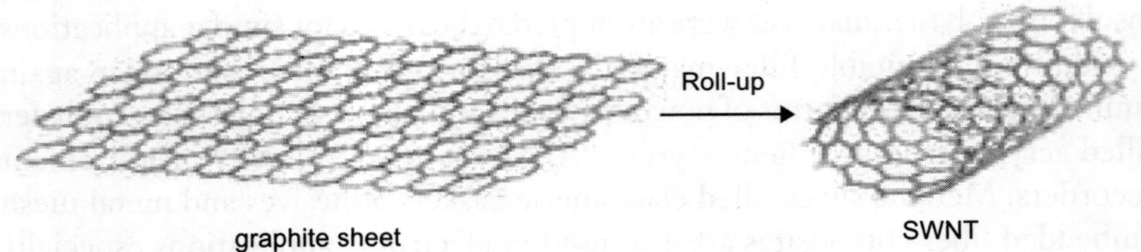

Fig. 11.1: Formation of SWNT.

Likewise, multiwall nanotubes are formed if two or more graphene sheets are rolled together, such that it results in a tubular structure with hollow concentric shells with spacing of 0.34 nm.

Now, if we consider any two arbitrary points O and A on the graphene sheet such that they coincide once the sheet is rolled to form a SWNT, then the vector joining the two points is defined as the chiral vector of the CNT, as explained in Figure 11.2. The line OA thus becomes the circumference of the SWNT cross section.

The CNT structures are often categorized on the basis of their chiral vector and chiral angle q. If $n = m$ and $q = 30°$, the resultant CNT is known as an **armchair nanotube**. If either n or $m = 0$ and $q = 0°$, it is referred as a **zigzag nanotube**. In all other cases, the CNTs are called **chiral nanotubes**.

In other words, CNT is a tube rolled along the chiral vector, the values of n and m determining the chirality or twist (q). The chirality in turn affects the conductance of the nanotube, its density, lattice structure and other properties.

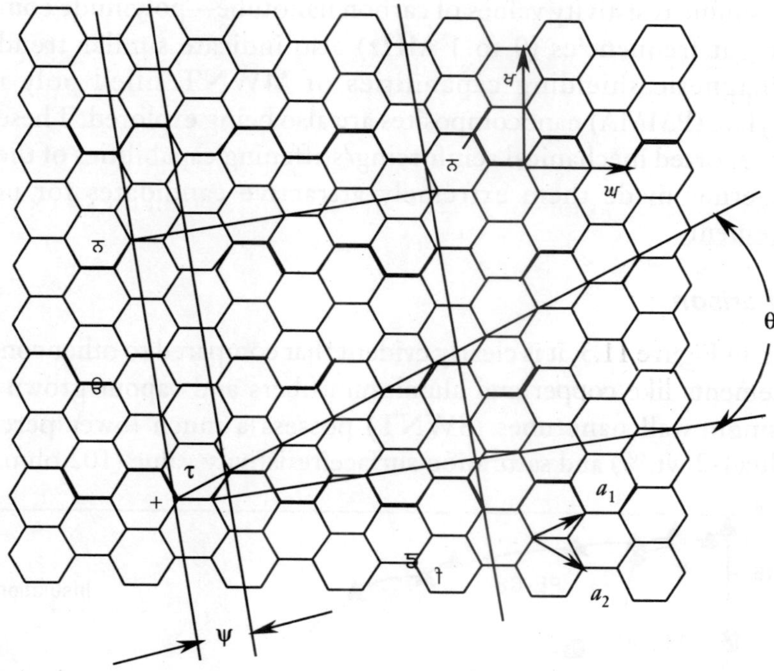

Fig. 11.2: Chiral Vector.

As mentioned above, not only structures, the electrical properties of the CNTs are also a function of their chirality. Depending on their chirality, CNTs are either semi-conducting or metallic7. The differences in conducting properties are caused by the different extent of p electron mismatch that results in a different band structure and band gap8. It was shown that a *(n, m)* nanotube behaves as a metallic conductor if *(n-m) = 3i*, where *i* is an integer. Otherwise, the CNTs behave as a semiconductor. All armchair CNTs are metallic. In theory, metallic CNTs can have electrical current density more than 1000 times stronger than metals like silver or copper. On the other hand, based on the exact values of *n*, *m* and *q*, the semi conducting CNT can have a band gap of either 0.01 ev or 1.0 ev (same order of conventional germanium/silicon semiconductors). This allows the designer to tune the materials for almost all levels of electrical conductivity. Generally, CNTs have very high electrical conductivity, much higher than those of conducting polymers.

Current State of Research

Reports on the effects of the length and aggregate size of multi wall nanotubes (MWNT) on the AC conductance of epoxy based composite reveals the percolation threshold for the fillers to be as low as 0.5 weight

%. The volume resistivity values of carbon nanotube—polyimide composites for different frequencies (0 to 1 MHz) also indicate similar trends. The electromagnetic shielding capabilities of MWNT filled poly methyl methacrylate (PMMA) nanocomposites are also being explored. These, along with the reported mechanical reinforcing/stiffening capabilities of the CNTs have together made them extremely attractive candidates for polymer reinforcement.

A Comparison

As shown in Figure 11.3, it is clearly evident that compared to other conductive reinforcements like copper and aluminium fibers and vapour grown carbon fibers, single wall nanotubes (SWNT) possess a much lower percolation threshold (1-2 wt.%) and saturation surface resistivity value (102 ohm/sq).

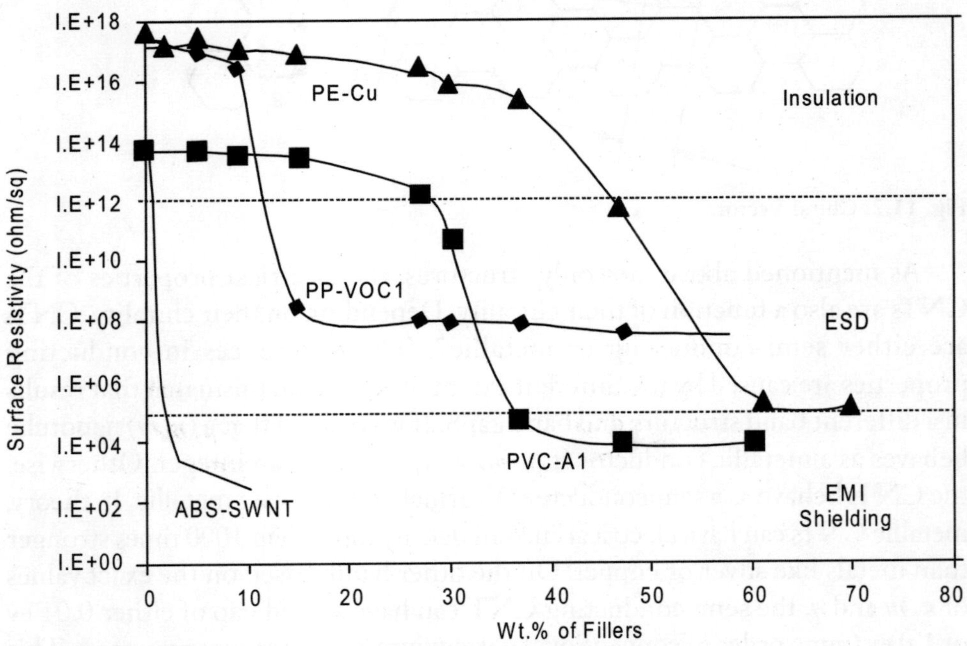

Fig. 11.3: Percolation phenomena in different conductive polymer composites.

Note: PE: Polyethylene; Cu: Copper fiber; PP: Polypropylene; VGCF: Vapour grown carbon fiber; PVC: Polyvinyl chloride; Al: Aluminium fiber; ABS: Acrylonitrile-butadiene-styrene; SWNT: Single wall nanotube.

Applications

The tendency of the CNTs to form ropes provides inherently very long conductive pathways even at ultra-low loadings, which may be exploited in

EMI/RFI shielding composites, gaskets, and other uses such as electrostatic dissipation (ESD), antistatic materials, transparent conductive coatings and radar absorbing materials (RAM) for stealth applications. The same properties also make these composites attractive for electronics packaging and interconnection applications like adhesives, potting compounds, coaxial cables and other types of connectors. In the long run, such multifunctional composite materials hold sufficient promise for applications in futuristic smart structures for aerospace and other critical applications.

Table 11.1: Techno-commercial Forecast For A 5"X 7" Esd Bag Of 2010, Made Out of Different Conductive Polymer Composites.

Conductive Filler	*Percolation Threshold*	*Cost (Cents)*
Metals (2005)	30%	20.0
Metals (2010)	30%	20.0
VGCF (2005)	11%	10.3
VGCF (2010)	5%	10.0
SWNT (2005)	1%	20.1
SWNT (2010)	0.5%	11.0

The table reflects the expected trend towards replacing the metal/carbon fiber filled conductive composites with their CNT counterparts. Two forces are primarily expected to drive this change:

❖ Lesser percolation threshold of CNTs, leading to lower filler loading & lightweight

❖ Decrease in cost, in view of the plunging prices of CNTs in international market

11.2 Fibre-Reinforced Plastic

A fibre-reinforced plastic (FRP) (also *fibre-reinforced polymer*) is a composite material comprising a polymer matrix reinforced with fibres. The fibers are usually fiberglass, carbon, or aramid, while the polymer is usually an epoxy, vinylester or polyester thermosetting plastic. FRPs are commonly used in the aerospace, automotive, marine, and construction industries.

11.3 Unsheathed Polymer Fibre

Core Material: Polymethyl Methacrylate
Cladding Material: Fluorinated polymer
Refractive Index: 1.49
Numerical Aperture: 0.5
Temperature Range: –55°C to +70°C
Attenuation: typically max. 200dB/km for 1mm fibre

Product Code	*Outer Diameter*	*Core Diameter*	*Reel Length*	*Approximate Weight (exc. packaging)*
CK10	0.25mm	240μm	6000m	0.06g/m
CK20	0.50mm	486μm	6000m	0.4g/m
CK30	0.75mm	735μm	2700m	0.9g/m
CK40	1.00mm	980μm	1500m	1.5g/m
CK60	1.50mm	1470μm	700m	3.2g/m
CK80	2.00mm	1960μm	250m	4.0g/m

Stranded fibre cables using 1mm and 1.5mm fibre can be made to order as can mixed diameter bundles.

11.4 Polymer Optical Fibres

Alex Argyros, Sydney University medallist in physics and doctoral candidate, won the Tertiary Student prize as part of the biannual Dupont awards for delivering a hollow-core polymer fibre that could transform the optical fibre industry.

"Up until now, optical fibres—used to reflect and carry light—have usually been made of glass or silica. This has restricted their use in some fields, such as medicine or surgery as thicker glass fibres are not flexible and they might shatter inside a persons body, making it very dangerous," explains Alex.

In contrast, hollow-core polymer optical fibres—made of a single material that transmits light even at wavelengths where the polymer is not transparent—are made from plastic, enabling them to be used for medical applications with a greater ease and less risk.

"Because it's plastic, it's safe to use internally, there is no danger of it shattering. It's also far more supple than glass, making it easy for surgeons to probe and see into hard to reach parts of the body and its cheaper, so you could make it more hygienic by manufacturing more disposable products—using it on one person and then throwing it out," he says.

Another unique aspect of this polymer optical fibre is its hollow core, which allows light to travel through without the material needing to be as transparent as you normally would in other types of optical fibres. This gives you more flexibility, allowing you to change the plastic to suit each specific application. For example, you can use a plastic that is stable at a higher temp, more flexible, cheaper or stronger. The material you use around the core is less important than before."

Alex and his research team from the University's Optical Fibre Technology Centre were awarded the Australasian Science Prize for 2005, which recognised their success in polymer optical fibres. Now, they are taking it further, with a new fabrication method which allows them to produce larger quantities at a reduced price.

Before, we started off with a large piece of plastic and drilled a large number of holes into it to create the hollow core. We drilled that many holes that it took us up to two weeks to create a preform—a large 7cm version of the fibre, which is stretched out to make the fibre. The new fabrication method allows us to create ones that are 40cm-50cm in 1 -2 days, making it cheaper and faster to produce, he explains. The new method simply involves stacking polymer tubes inside a larger tube to create the pattern of holes needed to guide the light.

These new fibres offer "many unique properties which will find important applications when they can be made more readily. This work is an important step in that direction. Furthermore, being easier to work, it will be possible to produce many unique structures in polymer fibres, which would be impractical in glasses," said Professor Simon Fleming, Technical Director of the Optical Fibre Technology Centre.

Glossary of Terms

Abrasive Wear
Wear caused by the continual contact, under pressure of hard particles in the resins against the barrel lining, screw and valve components. The abrasive particles may be fillers or reinforcements, such as fiberglass, calcium carbonate, powdered metals and others.

Achromatic
Colors without color, black and shades of gray.

Acrylic Sheet
A flat sheet of acrylic, variable size or thickness, often used as a work surface.

Active Center
Location of the unpaired electron on a free radical where reactions take place.

Additive
A substance compounded into a resin to modify its characteristics (i.e. intestates, stabilizers, plasticizers, flame retardants, etc.).

Adhesive Wear
Wear resulting from two metals rubbing against each other, such as the screw flight lands and valve rings coming into contact with the barrel lining during operation.

Alignment Layer
A layer and/or surface treatment applied to the boundary of a liquid crystal cell to induce a particular director orientation. For example, a layer of

polyimid buffed in one direction induces alignment parallel to the buffing direction, or a surfactant may be polymerized on a boundary surface to induce perpendicular alignment.

Alternating Copolymers

A copolymer comprising two species of monomeric units distributed in alternating sequence: the arrangement -ABABABAB- or (AB)n represents an alternating copolymer. Alternating copolymers are named:poly(A-alt-B)

Amorphous

Irregular; having no discernible order or shape. In the context of solids, the molecules are randomly arranged, as in glass, rather than periodically arranged, as in a crystalline material.

Amorphous

A term describing polymers having no crystalline structure.

Amorphous Polymers

A glass-like structure with tangled chains and no long-range order.

Amphiphilic

A molecule with a hydrophilic head and a hydrophobic tail. That is, a molecule that has one end which attracts water and one end which repels water.

Amphiphilic ('loving Both')

One end of an amphiphilic molecule is polar and hydrophilic (water loving) and the other is non-polar and hydrophobic (water hating). The hydrophilic ends of the molecule point outwards into the solution and the hydrophobic ends point inwards away from the water, so they tend to self-assemble in water. Amphiphilic materials are already widely used, but research into their use for drug delivery and ultrasound imaging is relatively new.

Analogous Colors

Colors next to each other on the color wheel.

Anisotropic

Having properties which vary depending on the direction of measurement. In liquid crystals, this is due to the alignment and the shape of the molecules.

Anneal

To heat an article to a predetermined temperature and slowly cool it to relieve stresses. The annealing of metal components may reduce their hardness.

Appliqué

Applied on top of.

Armature

A skeletal framework used as a structural support.

Aromatic

A compound containing a series of benzene (6 Carbon) rings; so named because many have a distinctive odor.

Atactic Polymer

A regular polymer, the molecules of which have equal numbers of the possible configurational base units in a random sequence distribution.

Auxetic Materials

Grow fatter when stretched and thinner when compressed – the opposite of a rubber band. Auxetic materials are resistant to impact, so they have possible uses as car bumpers, gaskets in engines, soundproofing and in bullet-proof vests.

Axial Ratio

The ratio of the length of the molecule to the diameter of the molecule.

Backbone

The main structure of a polymer onto which substituents are attached.

Ball Check Valve

A device mounted at the discharge end of an injection screw used to allow forward flow of plastic during recovery and no flow during the injection stroke. This valving is accomplished by positioning of an internal ball.

Bar

A small 2 oz unit of FIMO clay, or other polymer clay.

Barrel (Cylinder)

A cylindrical housing in which the screw rotates, including the integrally formed special inner surface material, or replaceable liner, if used. Also commonly referred to as a cylinder.

Barrel Shell (Backing Material)

The outer thick wall of the barrel made from metal backing material to provide strength and to hold the lining.

Barrier Flight

A secondary flight of reduced outside diameter and usually designed to separate melted polymer from solid polymer or to enhance melting by having the polymer pass over it.

Bayonet Adapter

A cylindrical shaped part with holding pins that threads into the thermocouple hole and retains a spring loaded thermocouple.

Bearing (HUB)

The portion of the screw immediately behind the flighted length which prevents the escape of material and provides a seal between the screw and the barrel.

Bell End

A flange at the discharge end of the barrel which provides added strength to withstand internal pressure.

Biaxial

Possess two directions along which monochromatic light vibrating in any plane will travel with the same velocity. The optic axis lies just between these directions.

Bilayer

A double layer of amphiphilic molecules, arranged such that either the nonpolar ends are on the inside screened by the polar ends or the polar ends are on the inside screened by the nonpolar ends, depending on whether the solvent is polar or nonpolar.

Bimetallic

A term used to indicate that a barrel is composed of two (or more) metals, commonly used to refer to barrels which have a centrifugally cast lining.

Biomaterial

A synthetic material used to replace part of a living system or to function in living tissue. A biomaterial is different from a biological material in that it is engineered rather than being naturally produced by a biological system.

Birefringence

Also called double refraction. The property of uniaxial anisotropic materials in which light propagates at different velocities, depending on its direction of polarization relative to the optic axis. A wave with polarization perpendicular to the optic axis will exhibit an "ordinary" index of refraction, n_o (this is often referred to as the ordinary ray). In contrast, a wave with polarization parallel to the optic axis exhibits an "extraordinary" index, n_e (the extraordinary ray). The ordinary index, n_o, is isotropic with respect to direction of propagation while the extraordinary, n_e, varies depending on the direction of propagation with a maximum value for light traveling perpendicular to the optic axis and, of course, polarized parallel to it. The difference face="Symbol">D n = n_e - n_o is also referred to as the birefringence or the optical anisotropy.

Block polymers

Polymers composed of two or more connected sequences (blocks) of homopolymers. In the simplest case, the AB diblock consists of two homopolymers, A and B, joined together.

Blister Ring

A raised portion of the root between flights of sufficient height and thickness to effect a shearing action of the polymer as it flows between the blister ring and the inside wall of the barrel.

Block Copolymers

A polymer comprising molecules in which there is a linear arrangement of blocks, a block being defined as a portion of a polymer molecule in which the monomeric units have at least one constitutional or configurational feature absent from the adjacent portions. Block copolymers are named: polyA-block-polyB.

Blow Molding

A method of fabrication in which a warm plastic parison (hollow tube) is placed between the two halves of a mold cavity and forced to assume the shape of that mold cavity by the use of air pressure. Pressurized air is introduced into the inside of the parison through a blow pin thereby forcing the plastic parison against the surface of the mold that defines the shape of the product.

Bond Orientational Order

Describes a line joining the centers of nearest-neighbor molecules without requiring a regular spacing along that line. Thus, a relatively long-range order with respect to the line of centers but only short range positional order along that line.

Bore

The inside diameter of the barrel which houses the screw.

Branched Polymer

A polymer with a chemical side chain extending from the main backbone.

Brayer

A roller tool that has a handle for leverage and increased pressure. Acrylic Brayers are suitable for work with FIMO because an artist can see the surface while rolling.

Breaker Plate (Extrusion)

A metal plate installed across the flow of the melt between the end of an extruder screw and the die adapter with openings through it such as holes or slots. It commonly supports a screen and also reduces any swirling in the melt flow before it reaches the die.

Breaker Plate Recess (Extrusion)

The internal counterbore at the discharge end of an extruder barrel that accepts the breaker plate and provides the seal and alignment of the die adapter.

Brick

A large 13 oz. unit of FIMO®.

Buffing

To give the inner glass surfaces of a liquid crystal cell a texture so as to align the liquid crystal molecules in a certain direction parallel to the surfaces.

Buffing Wheel

A machine driven wheel that has a muslin or flannel disk used for buffing. Also available are drill pads with muslin bonnets which also work in shining polymer surfaces.

Bull's Eye

A design of circles, increasing in diameter and graduating outward from the same midpoint.

Burnisher

A tool with a flat blunt wand used to smooth or shine an area using friction.

C Clamp Flange

A circular tapered flange at the discharge end of an extruder barrel used to hold the die adapter. A similar shaped flange on the die adapter and a "C" shaped clam accomplishes the closure.

Cane

A cylinder or length (any shape) of clay (or glass) wherein several rods or sheets of color have been placed together to form a design running lengthwise through the shape.

Ceramics

Are inorganic, non-metallic solids processed or used at high temperatures. A ceramic is made by combining metallic and non-metallic elements. Traditional ceramic products such as clay pots and chinaware are hard, porous and brittle. Modern ceramics are used to create bones and teeth, cutting tools or to conduct electricity.

Chain Axis

The straight line parallel to the directions of chain extension, connecting the centers of mass of successive blocks of chain units, each of which is contained within an identity period.

Chain Polymer

A polymer in which the repetition of units is linear. The monomers are linked end to end forming a single straight polymer.

Chain Polymerization

A chain reaction in which the growth of a polymer chain proceeds exclusively by reaction(s) between monomer(s) and reactive site(s) on the polymer chain, with regeneration of the reactive site(s) at the end of each growth step.

Channel

With the screw in the barrel, the space between the flights bounded by the root of the screw and the bore of the barrel.

Channel Area (Axial)

The cross-sectional areas of the channel measured in a plane through and containing the screw axis.

Channel Depth

The distance in a radial direction from the bore of the barrel to the root of the screw.

Channel Volume

The volume developed by the "axial area of screw channel" in one revolution about the screw axis. The location of measurement should be specified..

Channel Volume (Enclosed)

The volume of screw channel starting form the forward edge of the feed opening to the discharge end of the screw channel. Also called screw inventory.

Channel Width

The distance across the screw channel in a direction perpendicular to the flight measured at the periphery of the flight.

Check Ring

The cylindrically shaped component of a non-return valve that reciprocates in an axial direction. During injection, the ring shuts off against the surface of the rear seat of the valve preventing the melt from flowing backward toward the feed section of the screw. During screw recover, the ring rests against the front seat (or stud in a three piece valve) of the valve allowing melt to flow forward through the valve to the discharge end of the barrel.

Chiral Molecule

A molecule that is not identical to its mirror image. This gives a chiral substance its characteristic twisted shape, due to the fact that its molecules do not line up when combined.

Cholesteric Liquid Crystals

Also known as Chiral Nematic. Similar to the nematic phase, however, in the cholesteric phase, molecules in the different layers orient at a slight angle relative to each other (rather than parallel as in the nematic). Each consecutive molecule is rotated slightly relative to the one before it. Therefore, instead of the constant director of the nematic, the cholesteric director rotates helically throughout the sample. Many cholesterol esters exhibit this phase, hence the name cholesteric.

Cholesteric Mesophase

Nematic liquid crystals with chiral centers form in two dimensional nematic-like layers with directors in each layer twisted with respect to those above and below so that the directors form a continuous helix about the layer normal. Many cholesteric esters exhibit this phase, hence the name cholesteric. This mesophase exhibits circular dichroism and is optically active.

Chromophoric Groups

Chemical groups which have distinctive colors.

Circular Polarization

A condition caused by two waves whose electric field components are 90 degrees out of phase, causing an effective rotation of the electric field about an axis in direction of propagation.

Cis (configuration)

A polymer configuration in which adjacent bonds are coplanar and on the same side of the carbon-carbon double bond.

Clay Extruder

A hand operated tool that presses clay through an opening to extrude the clay in a continuous shape.

Clearance (Screw/Barrel)

The difference in the diameters of the screw and the bore (diametral clearance) or, more commonly, one-half the diametral clearance, referred to as radial clearance.

Coefficient Of Thermal Expansion

The fractional change in dimension (sometimes volume) specified of a material (plastic, metal or other materials) per a unit change in temperature.

Colligative Properties

Are those that depend on the number of species present rather than their kind.

Color Concentrate

A measured amount of die or pigment combined into a predetermined amount of resin, serving as a carner to for a color concentrate. The concentrate is then added to the bulk of the resin in measured quantity to achieve a predetermined color in the finished plastic product.

Columnar Phase

A liquid crystal phase characterized by disc-shaped molecules that tend to align themselves in vertical columns.

Combination

A type of termination reaction which involves a free radical joining with an active growing polymer to form a stable chain and end the growth.

Composites

Are formed by combining two or more materials that have quite different properties. The different materials combine to give the composite unique properties, but within the composite you can easily tell the different materials apart – they do not dissolve or blend into each other. Fibreglass is a composite material made from glass fibres which give it its strength and a flexible polymer resin (matrix) that binds the fibres together.

Compression Ratio (Channel Depth)

The factor obtained by dividing the channel depth in the feed section of the screw by the channel depth in the meter section (or the depth of the last complete flight). This Channel Depth ration is commonly referred to as the Compression Ratio. In constant lead screws this value approximates, but is greater than the volumetric compression ratio.

Compression Ratio (Volume)

The factor obtained by dividing the developed channel volume of the screw at the feed opening by the developed volume of the last full flight prior to discharge. This is not the commonly used definition of compression ratio.

Concentricity

The term to describe two circles or cylindrical shapes having a common center and common axis, such as the inside and outside diameters of the flighted surface and bearing surface of a screw. Deviation from concentricity is referred to as "Runout".

Conditioning

The preparation of polymer clay before use. Polymer clay must be kneaded or made pliable before use to allow the molecules in the clay to align properly for stretching.

Configuration

The geometrical arrangement in polymers arising from the order of atoms determined by chemical bonds.

Configurational Unit

A constitutional unit having one or more sites of defined stereisomerism.

Conformation

The geometrical arrangement in polymers arising from rotation about adjacent carbon-carbon single bonds.

Conformation Repeating Unit

The smallest unit of a polymer chain with a given conformation that is repeated along that chain through symmetry operations.

Conical Transition

Transition where the depth charge is accomplished with a conical shaped root surface (interrupted by the screw flights).

Contour Length

The maximum end-to-end distance of a linear polymer chain. For a single-strand polymer molecule, this usually means the end-to end distance of the chain extended to the all-trans conformation. For chains with complex structure, only an approximate value of the contour length may be accessible.

Convection Oven

An oven that heats with a continuous flow of air. This is particularly favorable when even and exact heating is desired.

Copolymer

Polymers that are derived from more than one species of monomer. Polymers having monomeric units differing in constitutional or configurational features but derived from a single monomer, are not regarded as copolymers.

Core

An internal hole extending longitudinally from the shank through a portion of the screw for the circulation of a heat transfer medium or installation of a heater.

Corrosive Wear

Wear caused by the attack of various acids on the surface of screws, barrels, valves and other processing components. Acids formed in the processing of plastic erode and pit the metal surfaces. The acids come from the polymers themselves or from flame retardants, forming and coupling agents.

Counterbore

The recessed area in the discharge end of an injection barrel which acts as a pilot to assure the concentric fitting and seal surface of the end cap to the barrel.

Crimp Bead

A tiny metal bead that clamps flat to hold wire by pressure.

Cross-linking

A process in which bonds are formed joining adjacent molecules. At low density, these bonds add to the elasticity of the polymer and at higher densities, eventually produce rigidity in the polymers.

Crystallinity

The presence of three-dimensional order on the level of atomic dimensions. In polymers, the range of order may be as small as about 2 nm in one (or more) crystallographic direction(s) and is usually below 50 nm in at least one direction. Polymer crystals frequently do not display the perfection that is usual for low-molecular mass substances. Polymer crystals that can be manipulated individually are often called polymer single crystals.

Crystallinity

A state of molecular structure in some polymers that denotes uniformity and compactness of the molecular chains forming the polymer.

Crystallization

The process of forming crystals from the melt or solution.

Cycle

The complete, repeating sequence of operations in a process or part of a process. In molding, cycling time is the elapsed time between a certain point in one cycle and the same point in the next.

Defects

A local break in the translational or orientational symmetry of the material.

Deformation

The condition where the director in a liquid crystalline material changes its orientation from one molecule to the next.

Degree Of Polymerization

The number of monomeric units in a molecule of a polymer. The DP determines many physical characteristics of the system.

Dendrimers

Highly branched molecules that have several layers of branching. These molecules exhibit a characteristic spherical shape.

Diastereomers

Stereoisomers that are not mirror images of each other. They have similar chemical properties but may have vastly different physical properties.

Dichroism

Refers to the selective difference in absorption between the two orthogonal components of the polarization state of light propagating in a given direction in an anisotropic medium. Generally applied to linearly polarized light. This is how a sheet polarizer works with one direction of polarization transmitted and the perpendicular directions absorbed. Circular dichroism describes the corresponding effect for circularly polarized light where one "handedness" of circularly polarized light is absorbed more strongly than the opposite "handedness." Solutions of chiral molecules will produce this effect as will cholesteric liquid crystals for wavelengths well removed from the pitch value.

Die

The metal orifice mounted to the discharge end of a extruder barrel through which the melt flows to form a desired extrudate. Commonly includes the die block and die adapter which holds the die block.

Die Adapter

Part of an extruder die which holds the die block.

Die Block

Part of an extruder die which contains the orifice through which the melt flows to form an extruded product.

Dipole

Two equal electric or magnetic charges of opposite sign, separated by a small distance. In the electric case, the dipole moment is given by the product of one charge and the distance of separation. Applies to charge and current distributions as well. In the electric case, a displacement of charge distribution produces a dipole moment, as in a molecule.

Director

The molecular direction of preferred orientation in liquid crystalline mesophases.

Disclination

Line defects arising from singularities in orientational order in a director field.

Discotic Liquid Crystal

The component disc-shaped molecules self-assemble in a way that resembles stacks of coins. The discs are stacked on top of each other to form columns, which in turn are packed on a two-dimensional, usually hexagonal, lattice.

Dislocations

Line defects arising from singularities in translational order in a crystalline lattice.

Dispersive Mixing

The mixing of a fluid (melted polymer) with a solid (unmelted polymer, pigment and agglomerates above a certa size, etc.) which exhibits a yield port. Dispersive mixing is involved in the final melting of a polymer or breaking down a pigment in the manufacture of a color concentrate.

Disproportionation

Disproportionation is a method of terminating chain polymerization which involves a free radical stripping a hydrogen atom from another free radical, so that one chain stabilizes with the hydrogen, and the other with a double bond.

Distributive Mixing

The mixing of a fluid (melted polymer) and other components (such as liquid color concentrate) where all components are fluids and do not exhibit a yield point. Distributive mixing is aimed at achieving thermal and color uniformity where no solids breakdown is required.

E-6000

Name brand of a silicate base glue, strong and water resistant.

Elastomers

A class of polymers that have some degree of cross linking and are rubbery. Elastomers possess memory, that is, they return to their original shape after a stress is applied.

Electro-optical Materials

Materials whose optical properties are changed under the application of an electric field.

Embossing Powder

Powders used in stamping design and when heat treated will become a shiny solid and rise slightly above the surface.

Emulsion

A mixture of two mutually insoluble liquids such that one is dispersed in the other in droplets which often cause the solution to be cloudy or translucent.

Enantiomers

Molecules which exist in two nonsuperimposable mirror images, analogous to human hands. Chiral molecules are perfect physical and chemical models of each other with the exception of their rotation of polarized light and those interactions that involve other chiral systems, such as chiral molecular recognition. A racemic mixture contains equal amounts of two enantiomers and thus produces no rotation of the plane of polarization of light.

Enantiotropic Liquid Crystal

Exhibit the liquid crystal state both when the temperature rises from the solid state side or when it falls from the liquid state. Monotropic liquid crystals exhibit the liquid crystalline state only when the temperature changes in one direction.

Endcap

The device that bolts to the discharge end of an injection barrel and adapts the injection nozzle.

Endcap Bolts

The bolts used to attach the endcap to the injection barrel.

Entanglement

A crossing of polymer chains, which, when subjected to a strain, remains intact and hence mechanically active.

Enthalpy

A measure of the heat energy in a system; thermodynamic systems tend to react in ways that decrease their enthalpy

Entropy

A measure of the disorder in a system; thermodynamic systems tend to react in ways that increase their entropy.

Epitaxial Growth

An ordered growth of material as it is deposited on an ordered substrate, in registry with the substrate (for example, preserving elements of the crystalline order of the substrate).

Extrudate

The stock or melt as it emerges from the discharge orifice (die) in a desired product form, such as film, pipe, coating on wire and others.

Eye Pin

A long straight rigid metal wire with a loop at one end used in connecting beads or jewelry findings.

Feed Housing

A separate components of an extruder barrel assembly which contains the feed openings, water cooling channels and, in some cases, a grooved interior lining. It is capable of withstanding high pressures (15,000 PSI or more), especially if it is grooved, and may incorporate a thermal barrier between it and the barrel to which it is attached. Also referred to as feed throat.

Feed Opening

A hole through the feed section of the barrel for the introduction of feed material into the barrel. Also referenced to as feed hole or feed port.

Feed Section (Screw)

The portion of the screw which receives the material to be processed and conveys it to the transition section of the screw. The feed section normally has a constant channel depth and constant root diameter.

Ferroelectric Material

One that produces domains of spontaneous polarization whose polar axis can be reversed in an electric field directed opposite to the total dipole moment of the lattice.

Fibers

Characterized by having a high length to diameter ratio, that is, the length is enormously greater than the diameter. As a result of the spinning process, most of the polymer molecules are oriented along the fiber axis. In regions where polymer molecules are perfectly oriented along the axis, crystals will be formed. Nearly perfect uniaxial orientation of the macromolecules represents the basic requirement for the preparation of strong polymer fibers.

Filler

An insert substance added to plastics for the purpose of improving physical properties or process ability, or to reduce the cost of material.

FIMO® Modeling Material

Polymer clay modeling material manufactured by Eberhard Faber, Gmbh, Germany.

Findings

Components used in making jewelry usually referring to the mechanical apparatus.

Flange

A short external section of the barrel, with a large diameter, through which bolt holes have been placed to either assist in mounting the barrel to the machine, or to which an end cap, die adapter or other member is fastened. It may also add strength to that section of the barrel.

Flight

The helical metal thread or raised portion of the screw.

Flight Cutback

The portion of the screw at the discharge end that is not flighted. This is normally included for calculations as part of the flight length.

Flight Depth

The distance in a radial direction from the land of the screw flight to the root.

Flight Face

The face of the flight extending from the root of the screw to the flight land. The rear face is the side toward the feed section and the front face is the side toward the meter end of the screw.

Flight Hard Surface

A screw flight having its periphery harder (or more wear resistant) than the base screw metal achieved by flame hardening, induction hardening, heat treating, depositing of hard facing metal, or other means.

Flight Land

The surface at the radial extremity of the flight constituting the periphery or outside diameter of the screw.

Flight Land Width

The distance across one flight land in a direction perpendicular to the flight. Commonly referred to as flight width.

Flight Lead

The distance in an axial direction from the center of a flight land to the center of the same flight land after one complete turn. This is not flight pitch in a mufti-flighted screw.

Flight Pitch

The distance in an axial direction from the center of a flight land to the corresponding point of an adjacent flight land.

Flighted Length

Overall axial length of the flighted portion of the screw, from the start of the feed pocket to the front end of the register or (in the case of extruder screws or smear head screws) the point where the mot diameter begins to decrease. Flighted length does not include valves.

Food Processor

A motorized chopper intended for food but also used to condition polymer clay.

Free Radical

A molecule with an unpaired electron, making it highly reactive.

Free Radical Polymerization

The synthesis of a polymer involving the chain reaction of free radicals with monomers.

Freedericksz Transition

The point at which a liquid crystal changes from an aligned to a deformed state under the influence of an external electric or magnetic field.

Fringed-micelle Model

A model of crystallinity in which the crystallized segments of a macromolecule belong predominantly to different crystals. Pure and Appl. chem, 1989, 61, 769 IUPAC Macromolecular Nomenclature for Crystalline polymers

Front Radius

The radius at the intersection of the front or pushing side of the flight and the screw root. Usually this radius is smaller than the rear radius and may change from one portion of the screw to another.

Front Seat (Valve)

A separate wear resistant component of a non-return valve that limits the forward travel of the check ring.

Gauche

A carbon-carbon bond in the main chain of a polymer that is rotated 120 degrees out of the plane. gauche(+) denotes a rotation of +120 degrees and gauche(-) denotes a rotation of -120 degrees—both as defined by the "right hand" rule.

Glass Transition Temperature

Tg, can be defined as the temperature at which the specific volume vs temperature plot has a change in slope. Chain motion and other local molecular motions are greatly reduced below this temperature, which is also dependant on the rate of cooling.

Glow In The Dark

Fluorescent or colors glowing under a black light.

Gold Leaf

A extremely thin sheet of imitation metal similar to a foil. Leafing comes in several colors, gold, silver and variegated.

Graduated Beads

Beads that are increasing or decreasing in size in equal proportions.

Grandjean Texture

A specific pattern of defects found in chiral nematic liquid crystals that is caused by the chiral nature of the crystal. Named for the French scientist F. Grandjean who worked with chiral nematic liquid crystals.

Grit

Coarseness of texture in sanding papers or powders.

Grooved Liner (Or Barrel)

A liner whose bore is provided with longitudinal grooves, usually for several diameters in the feed section only.

Guild

A group or association of kindred pursuits or having a common interest.

Harmony

In color, an acetic arrangement of parts to form a pleasing whole.

Head Pin

A straight wire pin with a nail head used in connecting components of jewelry, usually an end unit that dangles.

Heater Bands

The electrical heating elements mounted on or around the barrel, adapters, dies. nozzles! etc.

Heating Zone

A portion of the barrel length having independent temperature control of the heater bands.

Helix

The molecular conformation of a spiral nature, generated by regularly repeating rotations around the backbone bonds of a macromolecule.

Helix Angle

The angle of a screw flight at its periphery relative to a plane perpendicular to the screw axis. It is calculated as: Helix Angle = Arctangent (Lead) Pi x D)

Helix Residue

The smallest set of one or more successive configurational base units that generates the whole chain through helical symmetry.

Homeotropic Texture

A mesogen configuration in which the molecules are aligned normal to the boundary surfaces, as at the faces of a liquid crystal cell, as illustrated. Consequently the director will be normal to the surface. This orientation is generally obtained by the application of an electric field normal to the surface but can be achieved through surface treatment.

Homogeneous

An uniform structure or composition throughout. Having or possessing the same properties.

Homogeneous (planar) Texture

A mesogen configuration in which the molecules are aligned parallel to the boundary surfaces, as at the faces of a liquid crystal cell.

Homopolymer

A polymer that is constructed of identical monomers.

Hopper (Feed)

A funnel-shaped container mounted directly on the barrel over the feed opening to hold a reserve of material to be processed.

Hue

The actual color.

Hybridization

The combination of atomic orbitals on the same atom. (Example: sp2, the composite of the "s" and two "p" orbitals.)

Hydrophilic

"Water loving"; describes a molecule which is attracted to water.

Hydrophobic

"Water fearing"; describes a molecule which is repelled by water.

Hygroscopic

A term applied to polymers in their pellet or powder form indicating a tendency to absorb moisture from air.

Hypertext

Text that includes pointers to other text, pictures, movies, etc. "Clicking" on these links takes the reader to the object that it is pointed to. Links may point to documents on other computers connected on the Internet.

Indentity Period

The shortest distance along the chain axis for translation repetition of the chain structure. The chain identity period is usually denoted by c. Pure and Appl. chem, 1989, 61, 769 IUPAC Macromolecular Nomenclature for Crystalline polymers

Index Of Refraction

Ratio of the phase velocity of electromagnetic radiation in free space divided by the phase velocity in a given medium. It is greater than one except for rather special cases.

Initiation Reaction

The first step in chain polymerization. Initiation involves the formation of a free radical.

Initiator

A relatively unstable molecule that decomposes into a free radical. Used to "initiate" a polymer growth reaction.

Injection Pressure

The pressure of the injection screw against the melted plastics material (expressed in PSI), excluding any loss of pressure due to frictional drag of the screw, equal to the force acting upon the injection piston.

Injection Rate

The injection rate is the calculated rate of displacement of the screw (or plunger), expressed in cubic inches per second, computer at the injection pressure specified.

Inlay

A hard surface portion of a flight land not extending across the full flight width.

Inlay

A decorative inlaid pattern.

Intaglio

An engraving depressed below the surface of the material so that an impression from the design yields an image in relief.

Involute Transition

The transition where the depth change is accomplished with a root surface which remains parallel to the screw axis but changes its depths.

Isomer

A molecule which has an identical molecular formula to another molecule, but has a different structure.

Isomorphism

Statistical co-crystallization of units having different repeating constituents, which may either belong to the same copolymer chains (copolymer isomorphism) or originate from different homopolymer chains (homopolymer isomorphism). Isomorphism is a general term; in the strict sense, the crystal structure is essentially the same throughout the range of compositions.

Isotactic Polymer

A regular polymer, the molecules of which can be described in terms of only one species of configuration base unit (having chiral or prochiral atoms in the main chain) in a single sequential arrangement. In an isotactic polymer, the configurational repeating unit is identical with the configurational base unit.

Isotropic

Having properties that are the same regardless of the direction of measurement. In the isotropic state, all directions are indistinguishable from each other.

ITO

Indium Tin Oxide (indium oxide doped with tin). A transparent conductive material very commonly used for electrodes in displays or other applications which require conductivity along with light transmission.

Kevlar

Kevlar is a synthetic polymer fibre and is used when reduced weight, increased strength and long wear life are required. The tensile strength of Kevlar is more than three times greater than that of steel, and its density is less by a factor five, so it is ideal for making very strong, light, flexible structures. It has become a household name, being used for yacht sails, bullet-proof vests and in the aerospace industry.

L/D Ratio (Screw)

The ratio of the working flighted length of the screw (distance from the front edge of the feed opening to the forward end of the screw flight when the screw is in the forward position) to Its outside diameter. In practice, the ratio calculation is simplified to dividing the flighted length of the screw by its nominal diameter.

Lacquer

A surface coating that dries as a protectant film, gloss or mat, over an object.

Lamination

The process of making by uniting superposed layers of one or more materials.

Latex Gloves

Commercial or medical thin stretchable gloves used to protect hands and to prevent fingerprints from imprinting on a piece.

Lead (Screw)-

The distance measured parallel to the screw axis from one edge of the top of a screw flight to the same edge of the same flight after one complete turn. It is not the distance measured to the next screw flight in a multiflighted screw.

Liner

The wear resistant removable sleeve(s) in the barrel.

Lining

The internal wear resistant portion of an extrusion or injection barrel. The lining may be metallurgically bonded to the inside diameter of the barrel shell or be removable liners that are press or shrink fit into the barrel shell bore.

Lipophilic

Describes a molecule which is attracted to hydrocarbons.

Liquid Crystal

A thermodynamic stable phase characterized by anisotropy of properties without the existence of a three-dimensional crystal lattice, generally lying in the temperature range between the solid and isotropic liquid phase, hence the term mesophase.

Living Polymerization

Chain growth polymerizations proceeding in the absence of chain breaking terminations. This process can be used to produce essentially monodisperse polymers.

Loaf

A cane, long square shape that will eventually be sliced.

Log

A cylindrical cane, or shape of clay.

Lyotropic

Materials in which liquid crystalline properties appear induced by the presence of a solvent, with mesophases depending on solvent concentration, as well as temperature.

Macromolecule

A very large molecule. Many polymers are composed of hundreds of thousands of atoms, and are thus characterized as macromolecules.

Macroscopic Molecular Orientation

The property that allows the orientation of liquid crystals to be seen only on a large scale and cannot seen if a single layer is compared to any other single layer.

Main-chain Polymer Liquid Crystals
Polymer liquid crystals in which the mesogenic cores are a part of the main chain or backbone of the polymer.

Marbleizing
Mixing of two or more colors of clay to produce a streaked or marbled effect resembling stone.

Melt
Plastic material in a molten condition.

Melt Channel
The channel in a barrier screw designed to collect and convey forward the melted polymer.

Melting Transition Temperature
The temperature at which the substance loses its translational and orientational order, changing from a solid phase to a liquid phase.

Mesogen
Rigid rodlike or disclike molecules which are components of liquid crystalline materials.

Mesogenic Cores
The mesogenic core is the primitive structural unit of a polymer having the requisite anisotropic shape and attractive interactions to establish long range intermolecular order in its liquid phase. That is, it is what gives a polymer liquid crystal its liquid crystal properties.

Mesomorphic Substance
Another term for a liquid crystal material.

Mesophase
Equilibrium liquid crystalline phases formed with order less than three dimensional(like crystals) and mobility less than that of an isotropic liquid. Parallel orientation of the longitudinal molecular axes is common to all mesophases (long-range orientational order).

Metal Alloys
Metal mixtures with greater strength, hardness or malleability than their component metals. The ratio of each component determines the properties of the alloy. Modern alloys may be created by adding just a few per cent of another metal.

Metering Section

The portion of the screw at the discharge end with a constant flight depth and having a length of at least one turn of the flight.

Micelle

A spherical formation caused by an amphiphilic substance in a solution. The lyophilic end of the molecule tends to orient itself toward the outside of the sphere while the lyophobic end tends to orient itself toward the inside of the sphere.

Millefiori

Refers to a process developed by Venetian Glass makers where several rods of glass are fused together in a pattern (often star or floral). Pieces of the resulting designed cane were then applied to the surface of another piece of glass or used in lengths crosswise exposing the design. This application literally translated from Italian means "Thousand Flowers".

Mixing Section

An area of the screw with special geometry designed to enhance - distributive and/or dispersive mixing in the melted polymer.

Mokume Ganè

A Japanese metal smithing technique that is currently popular in applying to polymer clay. Layers of clay (and leaf) are altered by pushing up or sinking objects (or clay) into the layers to create positive/negative special relations. When the layers are then sliced horizontally, the "terrain"of the clay resembles rings or patterns as the artist had altered the layered clay.

Molar Mass

M, is the mass divided by the amount of the substance. Molar mass is usually expressed in g/mol or kg/mol units. The g/mol unit is recommended in polymer science.

Molecular Weight

The ratio of the average mass per formula unit of a substance to 1/12 of the mass of an atom of nuclide $_{12}$C.

Molecular Weight Distribution

Represents the distribution of number of polymer chains with a given length and molecular weight M.

Monochromatic

Colors made from tints and shades of the same hue.

Monochromatic Light

Light composed of only one specific wavelength.

Monolayer

A layer of amphiphilic molecules on the surface of a solvent, arranged such that the ends attracted to the solvent are in contact with it and the other ends point into the air.

Monomer

The simple chemical unit which, when many are joined together, form a polymer.

Monotropic

A type of material which exhibits the liquid crystalline state only when the temperature changes in one direction. This is generally a result of the liquid crystal phase being below the melting temperature of the solid, where the liquid crystal phase is only observed if the liquid is supercooled below the melting point.

Nematic Mesophase

Liquid crystals are characterized by long-range orientational order and the random disposition of the centers of gravity in individual molecules. Nematics may be characterized as the simplest spontaneously anisotropic liquids. Nematic phases are composed of rod-shaped molecular aggregates that are arranged with parallel but not lateral order.

Neoprene

The DuPont name for a synthetic rubber fabric made from polymers of chloroprene. While it can only stretch a little, it is very strong. Because of its durability, neoprene is used for many industrial and commercial applications.

Nitriding (Gas)

The surface hardening of certain alloy steels by heating the steel in an atmosphere of nitrogen (ammonia gas) at approximately 9500F. A very hard (70+Rc) case depth of .007" to .0 15" results which is wear resistant. Process is commonly used for barrel inside diameters, screws and valve components.

Nitriding (ION)

The surface hardening of certain alloy steels by heating the steel (to approximately 600 OF) in an atmosphere of hydrogen gas, adding an electrical charge to the steel and nitrogen gas, allowing a bombardment of the positively charged steel by hydrogen and nitrogen gas ions. This creates a hard (70+Rc) wear resistant, case hardness to a slightly greater and more uniform depth than gas nitriding with less distortion or contamination of the workpiece. Process is commonly used on screws and valve components.

Non-Return Valve

A device mounted at the discharge end of an injection screw that uses a sliding ring to allow flow of plastics in one direction only. On screw recovery it allows flow forward. On injection stroke it allows no flow and makes the screw perform like a plunger.

Nose Cone

The conical surface at the discharge end of an extrusion screw.

Nozzle

A device that threads into the endcap and adapts the nozzle tip to the endcap.

Nozzle Tip

A device that threads into the injection nozzle and adapts between the nozzle and the sprue bushing of the injection mold.

Number-average Molecular Weight

The number of molecules of length i, multiplied by the molecular weight of an ith unit divided by the total number of molecules present.

Onlay

The application of a piece of clay onto another piece of clay.

Opaque

Neither reflecting nor emitting light. Not transparent or translucent.

Optic Axis

In a uniaxial material, a single direction of propagation along which double refraction does not occur. The index of refraction for both polarization directions is n_o along this axis. This axis lies along the director for nematic liquid crystal.

Optical Activity

The plane of vibration of linearly polarized light rotates as it propagates through a medium. This rotation can occur in either a right or left handed direction. Since linearly polarized light can be regarded as the sum of right and left hand circularly polarized components, this optical activity corresponds to different indices of refraction for the two circular components(circular birefringence).

Order Parameter

S describes the orientational order of liquid crystalline material, allowing for the individual orientational deviation of the molecules from the director, which represents the average over the collection. Typically, S ranges from 0.3 to 0.9, depending on the temperature, with a value of unity for perfect order.

Orientational Order

Measure of the tendency of the molecules to align along the director on a long-range basis.

Overall Length (Screw)

The total length of the screw parallel to the screw axis. It includes the flighted length, the shank and any nose cone, but not the non-return valve.

Parchment Paper

Oven baking paper that is used when baking polymer clay to eliminate "shiny spots" on pieces.

Parison

The hollow tube of molten polymer which is pinched at one or both ends and inflated in a blows mold to make a hollow part.

Pasta Machine

A mechanical device intended to roll pasta but used in the polymer clay industry to condition, blend color, texturize and roll the clay into flat sheets of varying thickness.

Pearlescent

Having a shimmering quality, clay that includes a filler that reflects microscopic bits of light, soft sparkling inside the clay, not metallic.

Periodic Copolymers

Copolymers where the monomeric units appear in an ordered (for example, ABC) sequence. Periodic copolymers are named: poly(A-per-B-per-C)

Perturbations

Disturbances from an equilibrium condition.

Pilot (Barrels)

A cylindrical portion at the rear end of an extruder barrel used to locate the barrel with the feed throat.

Pilot (Screws)

An internal cylindrical surface at the front end of an injection screw used to accurately locate a non-return valve or other portion that attaches to the screw.

Pitch

The distance it takes for the director of a cholesteric liquid crystal to go through one complete rotation of 360 degrees.

Pitch

The distance measured parallel to the screw axis from one edge of the top of the screw flight to the same edge of the adjacent flight. In the case of a mufti-flighted screw the pitch is less than the lead.

Plasticizer

Material added to a polymer to improve its processability and/or flexibility. These are low molecular weight substances which, when mixed with a polymer, lower its glass transition temperature, T_g.

Plasticizer

The chemical agent that enables softness, or initial consistency to polymer clay.

Plasticizing Capacity

The maximum quantity of a specified plastic material that can be raised to a uniform and moldable temperature in a unit of time. This capacity is generally expressed in pounds per hour (or ounces per second) as calculated from the recovery rate.

Plastics

A large group of polymers that has properties between elastomers and fibers. As such, plastics have a wide range of properties such as flexibility and hardness and can be synthesized to have almost any combination of desired properties.

Pocket

A place where a screw flight is initiated, usually starting from a cylindrical area or another flight. A feed pocket exists on most screws and is located at the intersection of the bearing and the beginning of the flight.

Polarizability

Relates the induced electric dipole moment, p, of an atom or molecule to the local electric field it experiences as a = p / E_{local}, hence depending on the displacement by the field of the electronic charge from its equilibrium position in the atom or molecule.

Polarizer (linear)

A device, which in the transmission of electro-magnetic radiation, confines the vibration of the electric and magnetic field vectors to one plane.

Polydisperse Polymer System

A polydisperse polymer system is one in which there is a distribution of molecular weights present. A monodisperse system would have only one molecular weight present.

Polydispersity

The ratio of weight average molecular weight to number average molecular weight.

Polymer

A high-molecular weight organic compound, natural or synthetic' whose structure can be represented by small repeat units (monomers) forming chemical bonds with the same or other monomers. If two or more monomers are involved, a copolymer is obtained.

Polymer Clay

A modern modeling material composed primarily of PVC resin, plastizer and occasional filler material. A colorful clay like modeling compound that is cured at a temperature of 265°F, is permanent, and water resistant.

Polymer Liquid Crystals

Polymers that contain mesogen units and thus have liquid crystal properties.

Polymerization

The act of joining simple molecules(monomers)into giant ones (polymers) to form plastics, fibers, elastomers and non-structural resins.

Polymers

Polymers are large molecules that are made up of many units (monomers) linked together in a chain. There are naturally occurring polymers (eg, starch and DNA) and synthetic polymers (eg, nylon and silicone).

Polymers

Long chains of covalently bonded atoms.

Polymers Of Intrinsic Microporosity (PIMs)

Are organic polymers that are porous to certain substances and not to others, so they can filter out target molecules. They are useful in industrial processing, medical technologies and in the laboratory, and can supply clean drinking water.

Polymorphism

The ability to exist in more than one crystal structure.

Polyvinyl Chloride

Commonly known as PVC, a polymerized vinyl chloride compound, thermoplastic resin.

Positional Order

The extent to which the position of an average molecule or group of molecules shows translational symmetry.

Press Fit

An interference fit, characterized by constant bore pressure throughout, achieved by mechanically forcing the entering piece into the receiving piece by use of a press or similar machine. Barrel liners may be press fit into the prepared barrel shell.

Pressure Tap

A hole designed to accept a pressure transducer or pressure gauge.

Primary Crystallization

The first stage of crystallization, considered to be ended when most of the spherulite surfaces impinge on each other. In isothermal crystallization, primary crystallization is often described by the Avrami equation.

Prolate

Elongated at the poles.

Propagation (in Polymer Growth)

The irreversible repetitive addition of monomers to the growing chain.

Propagation Reaction

The middle step in chain polymerization where successive monomers are attached to the growing chain.

Protein

A large molecule composed of a linear sequence of amino acids. This linear sequence is a protein's primary structure. Short sequences within the protein molecule can interact to form regular folds (eg, alpha helix and beta pleated sheet) called the secondary structure. Further folding from interaction between sites in the secondary structure forms the tertiary structure of the protein. Proteins are essential to the structure and function of cells. They account for more than 50 per cent of the dry weight of most cells, and are involved in most cell processes. Examples of proteins include enzymes, collagen in tendons and ligaments and some hormones.

Pulver

Micro metallic powder used to coat polymer clay with a metallic sheen.

Pump Ratio

The ratio obtained by dividing the depth of the second metering section by the depth of the first metering section on a two-stage screw.

Pushing Side

The flight face of the screw flight that faces the discharge end and runs from the front radius to the top of the flight land. This surface is commonly close to being perpendicular to the axis of the screw.

PVC

Common term for polyvinyl chloride.

Raised Register

A register that has a larger diameter than the adjacent root diameter. This is sometimes supplied on injection screws having metering depths too deep to match the rear seat of a standard non-return valve.

Random Copolymers

Is a special case of a statistical copolymer. It is a statistical copolymer in which the probability of finding a given monomeric unit at any given site in the chain is independent of the nature of the neighboring units at that position (Bernoullian distribution). Random copolymers are named: poly(A-ran-B)

Rear Radius

The radius at the intersection of the rear or trailing side of the flight and the screw root. Usually this radius is larger than the front radius and may change from one portion of the screw to another.

Rear Seat

A flat, ring shaped portion of a non-return valve that abuts the front vertical face of an injection screw and seals the flow of plastic by contact with the rear conical shaped end of the check ring.

Recovery Rate

The weight (or volume) of a specified moldable material discharged from the screw per unit of time, when operating at 50% of injection capacity, as determined by SPI test procedure. The rating is normally expressed as ounces r cubic inches) per second.

Recrystallization

Reorganization proceeding through partial melting. Recrystallization is likely to result in an increase in the degree of crystallinity, or crystal perfection or both.

Reduction

The shrinking of a design pattern to a smaller scale. Usually refers to a cane that is created in a large scale and then reduced in proportion by various techniques.

Register

The cylindrical portion of an injection screw at the most forward end accurately machined to match the rear seat of the non-return valve.

Relaxation

A term used to mean all irreversible processes which bring a system back to equilibrium after it has been perturbed by some external force. For instance, if an electric field is applied to a fluid of polar materials a polarization will be induced, but this will disappear after the field is removed, because of the randomization of the molecular orientations produced by Brownian motions.

Relief

An area of the screw shank of lesser diameter than the outside diameter and located between the bearing and the spline or keyway.

Reorganization

The molecular process by which (i) amorphous or poorly ordered regions of a polymer specimen become incorporated into crystals, or (ii) a change to a more stable crystal structure takes place, or (iii) defects within the crystals decrease.

Reorganization

The molecular process by which (i) amorphous or poorly ordered regions of a polymer specimen become incorporated into crystals, or (ii) a change to a more stable crystal structure takes place, or (iii) defects within the crystals decrease.

Resilience

The capability of a strained body to recover its size and shape after deformation by comparative stress.

Resin

Any of a class of solid or semi-solid organic products of natural or synthetic origin, generally of high molecular weight with no definite melting point. In the broadest sense the term is used to designate any polymer that is a basic material for plastics. Most resins are polymers.

Resonance

A method of stabilizing a bond by delocalizing the electrons around the molecule.

Retainer

The larger part of a no n-return valve that threads into the injection screw. The forward portion retains the front seat or the sliding ring. The front end of the retainer is usually torpedo or conical shaped and usually is fluted.

Rockwell Hardness

A common method of expressing the degree of hardness of a material by testing its resistance to indentation, under pressure, by a diamond or steel ball. Results are expressed on various scales, the most common of which is the "C" scale, identified as Rc, which is used to compare the hardness of barrel linings, screw surfaces and valve components. A similar test, Brinnell hardness test, expresses results on a somewhat different scale and is identified as HB.

Root

The surface of the screw between the flights, usually a cylindrical or conical shape, of a diameter smaller than the outside diameter of the flights.

Rotational Symmetry

Consider the example of an unmarked billiard ball. Rotation of any amount about any axis through its center will take the ball into itself so it has complete rotational symmetry. However, the stitching on a baseball places several restrictions on the axes about which it can be rotated into itself.

Rotatory Power

Measure of the capability of a twisted liquid crystal layer to rotate the angle of incident polarized light.

Schlieren Texture

The texture that appears in the optical microscopy of nematic and related smectic C phases under crossed polarizers when the planarity of the phase is interrupted by defects. The schlieren, dark streaks or brushes, form in the liquid crystal, connecting the defect points. The dark streaks or brushes that are characteristic of this texture may also appear along disclinations in a liquid crystal.

Screen Pack

A woven metal screen or equivalent device installed across the flow of melt between the tip of an extruder screw and the die and supported by the breaker plate. It is used to strain out contaminants or to increase the back pressure or both.

Screw

A helically flighted shaft which rotates within the barrel to mechanically work and advance the material being processed.

Screw Axis

A reference line of infinite length drawn through the center at the rear of the shank and the center of the discharge end.

Screw Diameter

The dimension defined by the cross-section of the screw bounded by its flight lands. Although it is usually expressed as a nominal diameter (such as 3 1/2") its actual diameter could be 3.490" to 3.492".

Screw Speed

The number of screw revolutions per minute (RPM).

Sealing Ring

A round ring used in place of a breaker plate to seal between an extruder b barrel and the die adapter.

Secondary Crystallization

Crystallization occurring after primary crystallization, usually proceeding at a lower rate.

Self-assembly

The aggregation of molecular moieties into more ordered structures that are thermodynamically stable and involve noncovalent bonds. Crystallization is an example of such self-assembly. Self-assembly is used to build nanostrucures such as inorganic clusters and lattices, nanotubes and channels, host-guest complexes, monolayers, hydrogen-bonded networks and systems of intertwined molecules.

Semiconductor

Semiconductor is a material that conducts electricity at a level between an insulator and a conductor. The electrical properties of semiconductors can be controlled by adding small amounts of other atoms or impurities – called doping. Transistors made from semiconductors are used in all electronics including computers, mobile phones, calculators, CD and DVD players. Some semiconductors can also be made to emit light when exposed to an electric field, including diode lasers and light emitting diodes, or LEDs. Silicon is currently the most widely used semiconductor in computer chips and other electronic components.

Shank

The non-flighted portion of the screw at the drive end.

Shot

The yield from one complete Injection molding cycle, including the part, runner systems and flash.

Shot Capacity

The maximum volume of material which an injection molding machine can produce from one forward motion of the screw. Commonly expressed in ounces.

Shrink Fit

An interference fit, characterized by a constant bore pressure throughout, achieved by heating the larger receiving piece to allow the smaller mating piece to enter. In some cases, barrel liners are shrink fit into the prepared barrel shell.

Side Chain Polymer Liquid Crystal

A polymer liquid crystal in which the mesogens are attached to the side of the main chain and not a part of the main chain itself.

Smear Tip

A conical device used in place of a non-return valve at the discharge end of an injection screw. This smear tip is generally used with high viscosity, heat sensitive materials where a non-return valve would cause degradation.

Smectic Mesophase

The molecules organize themselves into layers. The smectic phases form a one dimensional periodic lattice in which the individual layers are two dimensional liquids. Now 12 different smectic phases have been identified.

Solids Channel

The channel in a barrier screw that is connected to the feed channel and is designed to contain and convey forward unmelted polymers.

Spacer

Flexible section of polymer chain between two mesogens or the mesogen and the backbone of a polymer.

Spintronics

Also known as 'spin-based electronics', is the science of using electrons to store data. It uses the charge on an electron as well as its 'spin' state to store 'qubits' of information. Spintronics may lead to a new way of calculating called quantum computing.

Spiral

A "Jelly Roll" design in a cane where sheets of clay are stacked and rolled to create a spiral design when sliced crossways.

Square Pitch

A screw flight where the pitch is equal to the nominal screw diameter, i.e., 3 1/2" diameter screw with a 3 1/2" lead

Stability

A measure of the reactivity of a given molecule.

Star-branching

A type of polymerization in which a branched polymer is formed as branches emanating from a single point.

Step-growth Polymerization

Often termed condensation, it is defined as a polymerization in which all molecular species in the system can react with each other ,and the growth of polymer chains proceeds by condensation reactions between molecules of all degrees of polymerization.

Stereoisomers

Isomers that differ only in the way their atoms are oriented.

Stereoregular Polymer

A regular polymer, the molecules of which can be described in terms of only one species of stereorepeating unit in a single sequential arrangement.

Stereorepeating Unit

A configurational repeating unit having defined configuration at all sites of stereoisomerism in the main chain of a polymer molecules.

Stereoselectivity

The chemical selectivity to polymerize only one stereoisomer (enantioselectivity and diastereoselectivity)

Stereospecific Polymerization

Polymerization in which a polymer of a specific tacticity is formed. A polymerization in which stereoisomerism present in the monomer is merely retained in the polymer is not to be regarded as stereospecific.

Steric Hindrance

A condition when the rotation of a given group is restricted due to the size of neighboring groups.

Stone Clay

Common name for "LaDoll" mineral base, air dry clay from Japan.

Stone Fimo®

Refers to the colors of a brand of FIMO® modeling material simulating stone.

Super Critical Fluids

A substance above its critical point on the temperature/pressure phase diagram. Above the critical point, the fluid is neither a gas nor a liquid but possesses properties of both. The viscosity of a supercritical fluid is at least one order of magnitude higher than the viscosity in the gaseous state, but is one or two orders of magnitude less than in the liquid state.

Superconductor

A substance that has no resistance to the flow of an electric current. Superconductors currently require very low temperatures to function. They can be used for energy storage, storing and retrieving digital information, medical imaging machines and friction free transport.

Surfactants

Surface active agents. Organic compounds consisting of two parts: a water-attracting (hydrophilic) portion and a water-resistant (hydrophobic) portion. Detergents may contain more than one kind of surfactant. The hydrophobic ends attach themselves to the soil particles or to the fabrics being washed while the hydrophilic ends are attracted to the water. The surfactant molecules surround the soil particles, break them up, force them away from the surface of the fabric, then suspend the soil particles in the wash water. Surfactants are classified by their ionic (electrical charge) properties in water.

Surging

A pronounced fluctuation in extruder output over a short period of time without a deliberate change in operating conditions.

Swap

Trade, an event popular amongst Internet members where a large group of artists simultaneously exchange pieces of a particular theme.

Swing Gate

A hinged device mounted at the discharge end of an extruder barrel and used to support the die adapter.

Swing Gate Flange

A flange at the discharge end of an extruder barrel using bolts to attach the die adapter to the barrel.

Symmetry

The invariance of some properties of the object being investigated with respect to all the transformations considered.

Syndiotactic Polymer

A regular polymer, the molecules of which can be described in terms of alternation of configurational base units that are enantiomeric. In the syndiotactic polymer, the configurational repeating units consist of two configurational base units that are enantiomeric.

T.I.R.

An abbreviation used to identify tolerances with respect to concentricity. Total indicator Reading is standard terminology for drafting and machinist work to describe the deviation in concentricity of a measured surface from a selected surface as shown on a dial Indicator. Also known as F.I.M. (Full Indicator Movement).

Tacticity

The orderliness of the succession of configurational repeating units in the main chain of a polymer molecule.

Teflon

A polymer of fluorinated ethylene with the chemical name polytetrafluoroethylene (PTFE). It was invented in 1938 and is one of the most slippery substances ever made.

Tensil Strength

Strength relating to capability of tension.

Tensile Strength

A measure of the ability of a polymer to withstand pulling stresses.

Termination Reaction

The final step in chain polymerization where the growth of a polymer is stopped.

Terpolymerization

Polymerization involving three types of monomers.

Thermocouple

A device, consisting of two dissimilar metallic conductors in contact. that produces an electrical current whose magnitude depends upon the temperature of the junction. The resulting temperature measurement feeds into a pyrometer to help monitor and control the temperature of a barrel or nozzle. The thermocouple is housed in a steel tubing adapter.

Thermocouple Hole

A machined hole in a barrel, nozzle or adapter designed accept the thermocouple adapter.

Thermoplastic

A material that will repeatedly soften when heated and harden when cooled. Typical thermoplastics are the styrenic polymers and copolymers, acrylics, cellulosics, polyethylene, polypropylene, vinyl's, nylons and various fluorocarbon materials.

Thermoplastics

Linear plastics of finite molecular weight that can be fabricated into complex shapes by melting and injection molding.

Thermoset

A material that changes as a result of a chemical reaction to heat and pressure, catalysts or ultra-violet light, from a soluble, fusible stage to an insoluble infusible and cured condition. Typical thermoset materials are unsaturated polyesters (BMC), aminos (melamine and urea), alkyds, epoxies and phenolics.

Thermosets

A type of plastic that must be cured, forming network-like structures that do not soften at high temperatures.

Thermotropic

Liquid crystal molecules which exhibit temperature dependent liquid crystalline behavior.

Threading

When a nemaparallelepipedtic liquid crystalline material shows a tangled, thread-like appearance when observed between crossed polaroids in the optical microscope.

Tint

Lightness of a color, the amount of white mixed with the hue.

Tissue Blade

An extremely sharp smooth blade borrowed from the medical industry that has the ability to slice thin precise layers.

Toggle Clasp

A clasp that has a "T" shape inserted into an opening (round or otherwise) to catch and hold two strands.

Topology

A branch of mathematics concerned with those properties of geometric configurations which remain unaltered under very general kinds of elastic deformations (transformations such as stretching or twisting) where length, angles, and shapes are changed.

Torsion

The rotation about a single bond which joins two atoms.

Trailing Side

The flight face of the screw that faces the feed end and runs from the rear radius to the top of the flight land.

Transducer

A device for transferring power generated in one system to another system. Various types of electronic transducers are used in connection with molding and extruding equipment to measure, monitor and/or control linear position, pressure, temperature, speed and other factors.

Transition Section

The portion of the screw between the feed section and the metering section in which the flight depth decreases in the direction of discharge. This section is sometimes referred to as the compression section.

Translational Order

A condition when molecules have some arrangement in space. Crystals have three degrees of translational order (each molecule is fixed in space with an x, y, and z coordinate) and liquids have no translational order.

Translucent

Shining or glowing through. Admitting and diffusing light so that objects beyond cannot be clearly distinguished. When referring to FIMO® - 00 Art Translucent, the sheerest clay which when used in thin layers or buffed, is almost clear.

Transparent

Having the property of transmitting light without appreciable scattering so that bodies beyond are visible. When referring to FIMO® - it is similar to porcelain, having a soft depth but not clear.

Twin Barrel

A dual cylindrical housing in which two extruder screws rotate side-by-side, including a nitrided inner surface or a centrifugally cast bimetallic lining for resistance to wear. There are two major types of twin barrels: one, which provides for the intermeshing of the two extruder screws and; a second, in which the two extruder screws do not intermesh. Intermeshing, counter-rotating twin screw extruders are commonly used in profile extrusion of thermally sensitive materials, such as PVC. Intermeshing and non-intermeshing, counter and co-rotating twin screw extruders are used in a variety of specialty polymer processing operations.

Two-Piece Barrel

A cylindrical housing in which the screw rotates, that is construction of two sections. One section, which extends the full length of the varrel, consists of an outer shell (typically constructed of annealed alloy steel) with a full outside barrel diameter over the majority of its length, then tapering abruptly to a much smaller diameter over the final one-fourth (approx.) of As length at the discharge end of the barrel. This section commonly has a centrifugally cast, bimetallic lining for resistance to wear. The second section, is an outer shell with full outside barrel diameter (construction of heat treated alloy steel), which is shrink fit over the smaller diameter of the first section and welded to the larger diameter of that section at its taper area. The second section provides added yield strength to the barrel at its discharge end.

Two-Stage Screw

A screw for a vented barrel consisting of a feed, transition and meter section, followed by a decompression section (or vent section) located under the vent to allow the escape of volatiles without vent bleed. The decompression section has a deep channel and is usually several diameters in length, followed by a second transition and meter section. Frequently, two-stage screws are longer (24 to 32:1 L/D) than standard screws, although two-stage screws may be as low as 20:1 L/D.

Uniaxial Materials

Possess only one direction along which monochromatic light vibrating in any plane will travel with the same velocity. This direction is know as the optic axis.

Unit Cell

The smallest, regularly repeating material portion contained in a parallelepiped from which a crystal is formed by parallel displacements in three dimensions. Unlike the case of low-molar mass substances, the unit cell of polymer crystals usually comprises only parts of the polymer molecules, and the regularity of the periodic repetition may be imperfect.

Value

The degree of lightness or darkness in a color.

Van Der Waal's Forces

Forces which act between molecules that are caused by small random fluctuations in the polarity of the molecules.

Varnish

The coating, covering, or applied to an object - or protective surface material.

Vent Bleed

The unplanned escape of melt through the vent during the operation of vented barrel processing.

Vent Deflector

A barrel with a vent which utilizes a two-stage screw to accomplish the removal of volatiles from the material being processed.

Vent Port

An opening through the barrel wall, intermediate in the injection or extrusion process, to permit the removal of air, water vapor and volatile matter from the material being processed.

Vent Stack (Chute)

A device surrounding a major portion of the vent deflector designed to prevent any material that might escape through the vent port during start up from collecting on the barrel, heater bands or wiring.

Viscosity

The internal resistance to flow existing between two liquid layers when they are moved relative to each other. This internal resistance is a result of interaction between liquid molecules in motion.

Vulcanization

A process by which a network of crosslinks is introduced into an elastomer to strengthen it.

Weight Average Molecular Weight

In the distribution of molecular weights represented in a polymer sample, the weight average is determined by summing the product of the number of molecules of length i and the square of the molecular weight of a molecule of length i and then dividing by the total weight.

Xerox Transfer

The transferring of an image printed on a Xerox copier and then applied to polymer clay in a technique that transfers the ink and subsequently the image to the clay.

Abbreviations

3-D	Three-Directional
4-D	Four Dimensional
A/C	Advanced Composite
AA	Allyl Alcohol
ABS	Acrylonitrile-Butadiene-Styrene
ADA	Adipic Acid
AES	Auger Electron Spectroscopy
AFPB	Asymmetric Four-Point Bend Test
AMPS	2-Acrylamido-2-Methylpropanesulfonic Acid
APC-1	Aromatic Polymer Composite
AQL	Acceptable Quality Level
ASTM	American Society for Testing and Materials
ATH	Alumina Trihydrate
ATLAS	Automated Tape Lay-Up System
ATR	Attenuated Total Reflectance
AWG	American Wire Gauge
BDMA	Benzyl Dimethyl Amine
BF3MEA	Boron Trifluoro-Mono-Ethylamine
BMC	Bulk Molding Compound
BMI	Bismaleimide Resins
BRDF	Bidirectional Reflectance Distribution Function
CAD	Computer Aided Design
CAI	Compression After Impact
CAM	Computer Aided Manufacturing
CASS	Copper Accelerated Salt Spray Test
CDP	Cresyl Diphenyl Phosphate
CF	Carbon Fibers
CFA	Composites Fabricators Association

CFMC	Carbon Fiber Management Council
CIP	Cold Isostatic Pressing
Circ	Circumferential Winding
CMC	Ceramic Matrix Composites
CP	Chemically Pure
CRT	Tension Testing Machine-Constant Rate of Traverse
CTL	Composite Tape Laying
CVD	Chemical Vapor Deposition
CVI	Chemical Vapor Infiltration
DAC	Diallyl Chlorendate
DAP	2,6-Diaminopyridine
DAP	Diallyl Phthalate
DCPD	Poly(dicyclopentadiene)
DDA	Dynamic Dielectric Analysis
DDS	Diamino Diphenyl Sulfone
DDTA	Derivative Differential Thermal Analysis
DETA	Diethylene Triamine
DICY	Dicyandiamide
DL	Dead Load
DMAC	Dimethyl Acetamide
DMC	Dough Molding Compound
DMF	Dimethylformamide
DMG	Dimethyl Glutarate
DMS	Dynamic Mechanical Spectroscopy
DMT	Dimethyl Terephthalate
DP	Degree of Polymerization
DPA	Diphenylamine
DRIFT	Diffuse Reflectance-FTIR
DSC	Differential Scanning Calorimeter
DTA	Differential Thermal Analysis
DTUL	Deflection Temperature Under Load
DVB	Divinylbenzene
EBC	Electron Beam Cutting
EBM	Electron Beam Machining
EBW	Electron Beam Welding
ECDG	Electrochemical Discharge Grinding
ECG	Electrochemical Grinding
ECH	Electrochemical Honing

ECM	Electrochemical Machining
ED	Edge Distance
EDM	Electric Discharge Machining
EMF	Electromotive Force Series
EMI	2-Ethyl-4-Methylimidazole
EMI	Electromagnetic Interference
EMM	Electromechanical Machining
EOL	End-of-Life
ESC	Environmental Stress Cracking
ESD	Electrostatic Discharge
FEP	Fluorinated Ethylene Propylene
FRM	Fiber-Reinforced Metal
FRS	Fiber-Reinforced Superalloy
FTIR	Fourier Transform Infrared Spectroscopy
GIM	Gas Injection Molding
GPC	Gel Permeation Chromatography
GRP	Glass-Reinforced Plastic
HAZ	Heat-Affected Zone
HDM	Hydrodynamic Machining
HHPA	Hexahydrophthalic Anhydride
HIP	Hot Isostatic Pressing
HMTA	Hexamethylenetetramine
HPLC	High Performance Liquid Chromatography
HPW	Hot Pressure Welding
IDF	Insertion and Deletion of Fibers
IGA	Intergranular Attack
IM	Injection Molding
IPN	Interpenetrating Polymer Networks
IRP	Infrared Polymerization Index
ISS	Ion Scattering Spectrometry
IVD	Ion Vapor Deposition
K&S	Kraemer-Sarnow Method
KHN	Knoop Hardness Number
LBC	Laser Beam Cutting
LBM	Laser Beam Machining

LCP	Liquid Crystal Polymers
LDC	Liquid Dynamic Compaction
LDPE	Low Density Polyethylene
LFL	Lower Flammability Limit
LIM	Liquid Injection Molding
LIMS	Laser Ionization Mass Spectrometry
LMC	Least Material Condition
LSD	Least Significant Difference-5% Level
MAC	Maximum Allowable Concentration
MBBPA	Methyliminobispropylamine
MBK	Methyl Butyl Ketone
MBR	Multiple Internal Reflection
MCF	Metal Coated Fibers
MDA	Methylene Dianiline
MEK	Methyl Ethyl Ketone
MEKP	Methyl Ethyl Ketone Peroxide
MMC	Metal Matrix Composites
MPDA	Metaphenylene Diamine
MRI	Magnetic Resonance Imaging
MTHPA	Methyltetrahydrophthalic Anhydride
MVP	Moisture Vapor Permeability
MVT	Moisture Vapor Transmission
MYS	Microyield Strength
NDT	Nondestructive Test
NMA	Nadic Methyl Anhydride
NNS	Near-Net Shaping
NODTM	Tri(n-Octyl n-Decyl) Trimellitate
NPG	Neopentyl Glycol
NVM	Nonvolatile Matter
OFC	Oxyfuel Gas Cutting
OFW	Oxyfuel Gas Welding
OHW	Oxyhydrogen Welding
OOR	Out-of-Roundness
P/M	Powder Metallurgy
PAM	Plasma Arc Machining
PAPA	Polyazelaic Polyanhydride

PAS	Photoacoustic Spectroscopy
PAT	Polyaminotriazoles
PAW	Plasma Arc Welding
PBI	Polybenzimidazoles
PBT	Polybenzothiazole
PBT, PBZ	Polyphenylene Benzobisthiazole
PC	Polycarbonate Resins
PCB	Printed Circuit Boards
PCL	Polycaprolactone
PCTFE	Polychlorotrifluoroethylene
PE	Pentaerythritol
PE	Polyethylene
PEEK	Polyetherether Ketone
PEG	Polyethylene Glycol
PEI	Polyetherimide
PEK	Polyetherketone
PES	Polyethersulfone
PET	Polyethylene Terephthalate
PHN	Pfund Hardness Number
PI	Polyimides
PMDA	Pyromellitic Dianhydride
POM	Polyoxymethylene
POM	Polyoxymethylenes
PP	Polypropylene
PPO	Polyphenylene Oxide
PPS	Polyphenylene Sulfide
PPSU	Polyphenylenesulphone
PSA	Pressure-Sensitive Adhesives
PSO	Polysulfones
PTDQ	Trimethyl Dihydtoxyquinoline
PTFE	Polytetrafluorethylene
PVA	Polyvinyl Alcohol
PVC	Polyvinyl Chloride
PVF	Polyvinyl Fluoride
PVI	Polyvinyl Isobutyl Ether
RA	Reflection-Absorption
RESS	Rapid Expansion of Supercritical Solutions
RH	Relative Humidity
RIM	Reaction Injection Molding

RP	Reinforced Plastic
RPLC	Reverse Phase Liquid Chromatography
RRIM	Reinforced Reaction Injection Molding
RS	Rapid Solidification
RT	Radiographic Testing
RTF	Reinforced Thermoplastics
RTM	Resin Transfer Molding
RW	Resistance Welding
s/s	Shipset
SACMA	Suppliers of Advanced Composite Materials Association
SANS	Small-Angle Neutron Scattering
SAXS	Small-Angle X-Ray Scattering
SDAP	Super-Diallyl Phthalate
SHS	Self-Propagating High Temperature Syntheses
SIMS	Secondary Ion Mass Spectrometer
SMA	Styrene Maleic Anhydride
SMAW	Shielded Metal Arc Welding
SMC	Sheet Molding Compound
SRM	SACMA Recommended Methods
STEB	Standard Test and Evaluation Bottle
STEM	Scanning Transmission Electron Microscope
TAC	Triallyl Cyanurate
TAP	Triallyl Phosphate
TBT	Tetrabutyl Titanate
TCP	Tricresyl Phosphate
TDA	Toluene-2,4-Diamine
TDI	Toluene-2,4-Diisocyanate
TEP	Triethyl Phosphate
TFA	Tetrahydrofurfuryl Alcohol
TFE	Tetrafluoroethylene
TGA	Thermogravimetric Analysis
THF	Tetrahydrofuran
TLV	Threshold Limit Values
TMEDA	Tetramethylethylenediamine
TML	Total Mass Loss
TMPD	2,2,4-Trimethyl-l,3-Pentanediol
TOF	Trioctyl Phosphate
TP	Thermoplastic

TPA	Terephthalic Acid
TPI	Turns per Inch
TRS	Time Resolved Spectroscopy
TSC	Thermal Stress Cracking
TTB	2,2' ,6,6' -Tetrabromo-3,3' ,5,5' -Tetramethyl-4,4' Dihydroxydiphenyl
UF	Urea-Formaldehyde Resin
UP	Unsaturated Polyester
UPS	Ultraviolet Photoelectron Spectroscopy
USW	Ultrasonic Welding
UV	Ultraviolet
UV/VIS	Ultraviolet/Visible
WAXS	Wide Angle X-ray Scattering
WFS	Wet Flexural Strength
WVT	Water Vapor Transmission
XPS	Photoelectric Spectroscopy

Bibliography

Adawi et al, *Organic Electronics*, 7 (2006) 222-228.

Applications of Thermoplastics in Electronics, Electronic Packaging & Prodn., July 1992, pp. 44-49.

Bai, J.B., Allaoui, A., "Effect of the length and the aggregate size of MWNTs on the improvement efficiency of the mechanical and electrical properties of nanocomposites-experimental investigation", Composites Part A: applied science and manufacturing, 34 (2003) pp 689-694

Boroumand et al, *Nano Letters* 5 (2005) 67-71.

D. Reith, H. Meyer et F. Müller-Plathe, Comp. Phys. Comm. 148, 299 (2002).

D. Reith, H. Meyer et F. Müller-Plathe, Macromolecules 34, 2335 (2001).

Dynamic Study of Polymer Films by Friction Force Microscopy With Continuously Varying Load", Xiaoping Wang, O.K.C. Tsui, Xudong Xiao, *Langmuir*, 18(18), 7066-7072 (2002).

Dynamics of Polymers Confined in Thin Films, O. K. C. Tsui, Binyang Du, In Recent Advances of Polymer Science Overseas, eds. Tianbai He and Hangjie Hu, Chemical Industry Publishing Co., Beijing, Chapter 16, 246-263 (2001).

Effect of Chain Ends and Chain Entanglement on Glass Transition Temperature of Polymer Thin Films, O. K. C. Tsui, H. F. Zhang, Macromolecules, 34, 9139-9142 (2001).

Effect of Interfacial Interactions on the Glass Transition of Polymer Thin Films, O. K. C. Tsui, T.P. Russell, C.J. Hawker, Macromolecules, 34, 5535-5539 (2001).

Effect of Interfacial Interactions on the Glass Transition of Polymer Thin Films, O. K. C. Tsui, T. P. Russell, C. J. Hawker, *Slow Dynamics in Complex System: 3rd International Symposium, Sendai, Japan, Nov. 2003., AIP Conference Proceedings Vol. 708*, M. Tokuyama and I. Oppenheim, eds., pp. 598-600 (2004).

Effect of Low Surface Energy Chain Ends on the Glass Transition Temperature of Polymer Thin Films, Fengchao Xie, H. F. Zhang, Fuk Kay Lee, Binyang Du, Ophelia K. C. Tsui, Y. Yokoe, K. Tanaka, A. Takahara, T. Kajiyama, Tianbai He, Macromolecules, 35, 1491-1492 (2002).

G. Strobl, Eur. Phys. J. E 3, 165 (2000).

H. Meyer et F. Müller-Plathe, J. Chem. Phys. 115, 7807 (2001).

H. Meyer et F. Müller-Plathe, Macromolecules 35, 1241 (2002).

Hobson et al, *Appl. Phys. Lett.* 81 (2002) 3519-3521.

In: *Polymer Crystallization: Observations, Concepts and Interpretation*, edited by J.U. Sommer et G. Reiter, Lecture Notes in Physics, Vol. 606 (Springer, Berlin-Heidelberg, 2003).

J. Chem. Theo. Comput. 2, 616 (2006)

J. F. Steffe (1996) *Rheological Methods in Food Process Engineering* 2nd ed.page 1.

J.P.K. Doye and D. Frenkel, J. Chem. Phys. 109, 10033-10041 (1998)

J.P.K. Doye and D. Frenkel, J. Chem. Phys. 110, 2692-2702 (1999)

J.P.K. Doye and D. Frenkel, J. Chem. Phys. 110, 7073-7086 (1999)

J.P.K. Doye and D. Frenkel, Phys. Rev. Lett. 81, 2160-2163 (1998)

J.P.K. Doye and D. Frenkel, Polymer 41, 1519-1528 (2000)

Kim. H. M., Kim. K., Lee. C. Y., Joo. J., Cho. S. J., Yoon. H. S., Pejacovic. D. A., Yoo. J., Epstein. A. J., "Electrical conductivity and electromagnetic interference shielding of multiwalled carbon nanotubes composites containing Fe catalyst", Applied Physics Letter, Vol. 84, No.4 (26 January 2004) pp 589-591

Kumagai, Naoichi; Sadao Sasajima, Hidebumi Ito (15 February 1978). "Long-term Creep of Rocks: Results with Large Specimens Obtained in about 20 Years and Those with Small Specimens in about 3 Years". *Journal of the Society of Materials Science (Japan)* 27 (293): 157–161. Japan Energy Society.

Lidzey et al, *Nature* 396 (1998) 53-55.

Lidzey et al, *Organic Electronics* 6 (2005) 221-228.

Olivero, A., Radford, D. W., "Integrating EMI shielding into composite structure", SAMPE Journal, Vol. 33, No. 1, January/February 1997, pp. 51-57.

P. Avouris, "Carbon nanotube electronics", *Chemical Physics* 281, (2002) pp. 429-445.

Sankaran S., Dasgupta S., Ravi Sekhar K., Jagdish Kumar M. N., "Thermosetting Polymer Composites for EMI Shielding Applications", Proceedings of International Conference on Electro-Magnetic Interference and Compatibility - INCEMIC 2006, (23-24 Feb 2006), pp. 01-06.

Sankaran. S., Dasgupta. S., Ravi Sekhar. K., Ghosh. C., "Carbon nanotubes reinforced syntactic composite foams", Proceedings of ISAMPE National

Conference on Composites - INCCOM IV, (09-10 Dec 2005), pp. 82-82XII.

Studying Surface Glass-to-Rubber Transition Using Atomic Force Microscopic Adhesion Measurements, O.K.C. Tsui, X.P. Wang, Jacob Y.L. Ho, T.K. Ng, Xudong Xiao, Macromolecules, 33, 4198-4204 (2000).

Suppressing EMI by Proper Gasketing, Electronic Packaging & Production, July 1992, pp 52-54

Surface Viscoelasticity Studies of Ultrathin Polymer Films Using Atomic Force Microscopic Adhesion Measurements, X. P. Wang, Xudong Xiao, O. K. C. Tsui, Macromolecules 34, 4180-4185 (2001).

T. Vettorel, H. Meyer *Coarse-graining of short polyethylene chains for studying polymer crystallization*

T. Vettorel, H. Meyer, J. Baschnagel, M. Fuchs *Structural Properties of Crystallizable Polymer Melts: Intrachain and Interchain Correlation Functions*

Toshio, O., Yuichi, I., Takashi. I., Rikio. Y., "Characterization of multi-walled carbon nanotubes/phenylethynyl terminated polyimide composite", Composites Part A: applied science and manufacturing, 35 (2004) pp. 67-74.

Tsotra, P., Friedrich, K., "Electrical and mechanical properties of functionally graded epoxy-resin/carbon fibre composites", Composites Part A: applied science and manufacturing, 34 (2003) pp. 75-82.

W. R. Schowalter (1978), Mechanics of Non-Newtonian Fluids Pergamon.

Wenus *et al.*, *Appl. Phys. Lett.* 85 (2004) 5848-5850.

Wenus *et al.*, *Phys. Rev. B.* (2006).

Index